SOTHIS PRESS

Sothis Press aims to completely reimagine the public domain book with an emphasis on universal access to captivatingly designed print editions. We believe that technology will enable every single book ever published to be perpetually accessible in beautiful print form. Sothis is beauty in ubiquity.

www.sothis.press
1 Waterhouse Square, London, EC1N 2ST

The Modern House-Carpenter's Companion and Builder's Guide

Being a Hand-Book for Workmen, and a Manual of Reference for Contractors and Builders

W. A. Sylvester

SOTHIS PRESS

ISBN 978-1-0365-0310-9

This edition is published by
Sothis Press © 2023. The company
is registered in England and Wales.
The Company Number is 14900389.

THE MODERN HOUSE-CARPENTER'S COMPANION AND BUILDER'S GUIDE:

Being a Hand-Book for Workmen, and a Manual of Reference for Contractors and Builders;

GIVING

RULES FOR GETTING THE LENGTHS AND FINDING THE BEVELS FOR RAFTERS FOR PITCH, HIP, AND VALLEY ROOFS; THE CONSTRUCTION OF FRENCH AND MANSARD ROOFS; SEVERAL FORMS OF TRUSSES, STAIRS, SPLAYED AND CIRCULAR WORK, ETC.; TABLE OF BRACES, SIZES AND WEIGHTS OF WINDOW SASH, AND FRAMES FOR THE SAME; TABLE OF BOARD, PLANK, AND SCANTLING MEASURE, ETC.

ALSO INFORMATION FOR THE CONVENIENCE OF BUILDERS AND CONTRACTORS IN MAKING ESTIMATES; EXPLANATIONS OF THE USES OF THE VARIOUS MARKINGS ON RULES AND SQUARES; THE SLIDE-RULE, AND HOW TO USE IT; STRENGTH OF MATERIALS; AND RULES FOR ESTIMATING THE SIZES OF BEAMS, COLUMNS, ETC.; AND SEVERAL PLANS FOR HOUSES.

Making the most comprehensive work for the price yet published.

BY W. A. SYLVESTER.

FORTY-FIVE FULL-PAGE PLATES.

FIFTH THOUSAND, ENLARGED.

BOSTON:
CUPPLES, UPHAM & CO.,
Old Corner Bookstore.
1884.

INTRODUCTION.

MANY workmen carry with them a small memorandum-book, in which they note down the rules for doing various jobs of work, and items to assist them in estimating. This book was begun in this way: the author adding rules and other information from time to time, till he conceived the idea of having it published, with additions and illustrations, for the convenience of workmen.

There are many large and valuable works on carpentry, but there are comparatively few workmen who feel able to pay from five dollars to fifteen dollars for a work of this kind: and, even if he owns one, he cannot carry it with him, it being too bulky; so that oftentimes when he needs such assistance as his book gives, he is obliged to do without.

Then, again, this book contains much that is of great use to the workman which larger works do not give; they going deeper into geometrical staircases, hand-rails, groined arches, and complicated work generally, which not one workman in fifty ever has occasion to do. We have given descriptions and illustrations sufficient to enable any ordinary workman to plan and construct a flight of straight or winding stairs.

Great care has been used to give rules and items to assist in estimating that are correct and reliable. The illustrations and descriptions of the markings on rules and squares, and especially the article on the slide-rule, will be appreciated by every workman. The article on strength of materials, and rules for estimating the sizes of timbers, posts, and beams, is very valuable. At the end of the book will be found a very complete glossary of architectural terms and definitions. This book has been published in a small form, so as to be convenient to carry in the coat-pocket; and the very low price will enable every workman to own one; and we hope and trust that the workmen generally will find it a help and convenience.

THE AUTHOR.

THE

MODERN HOUSE-CARPENTER'S COMPANION

AND

BUILDER'S GUIDE.

Plate 1. Fig. 1. *To bisect a given line.*—Let *a b* be the given line. With a radius somewhat more than half of the length of this line, and using the points *a* and *b* for centres, describe arcs intersecting above and below the given line, through which points of intersection draw the line *c d*.

Plate 1. Fig. 2. *To bisect an angle.*—Let *a b c* be the given angle. With *b* as a centre, and with any radius, describe the arc *d e*. Then, using *d* and *e* for centres, and with a radius somewhat more than half of the length of *d e*, describe arcs intersecting at *f*. Then draw a line from *b* through the intersection at *f*.

Plate 1. Fig. 3. *Given a tangent to a circle, to find the exact point of contact.*—Let *a b* be a tangent to the circle, the centre of which is at *c*. Draw a line from *c* to any point on the tangent, as at *d*. From *e*, the centre of this line, and with a radius equal to *c e*, describe an arc. The point where this arc crosses the tangent at *f* is the exact point of contact.

Plate 2. Fig. 4. *To describe an ellipse with a cord or thread.*—Draw the line *a b* representing the length of the required ellipse. Bisect this line (see Plate 1, Fig. 1); which gives the line *c d*, the length of which must be equal to the width of the required ellipse. With a pair of compasses, take the length of *a e*. Then, with *c* as a centre, describe arcs intersecting the line *a b* at *f* and at *g*: at each of these three places, *f*, *g*, and *c*, stick in a pin. Now pass a piece of cord or thread around these pins, draw it taut, and tie it. Now remove the pin from *c*, and, holding a pencil in the bight of the cord, draw it around through *c*, *b*, *d*, and *a*, keeping the thread at a uniform tension. A notch made in the side of the pencil-lead, near the point, will prevent the thread from slipping off.

A wire thread about the size of No. 40 or 60 linen thread would be better to use, as it will not stretch. It would be a good plan for the work-man to keep about twenty-five or thirty feet of it in his chest, rolled up on a spool, the same as a

Plate 1.

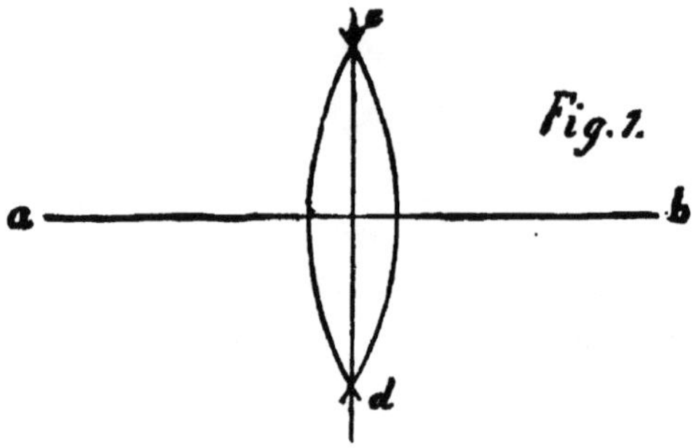

Fig. 1.

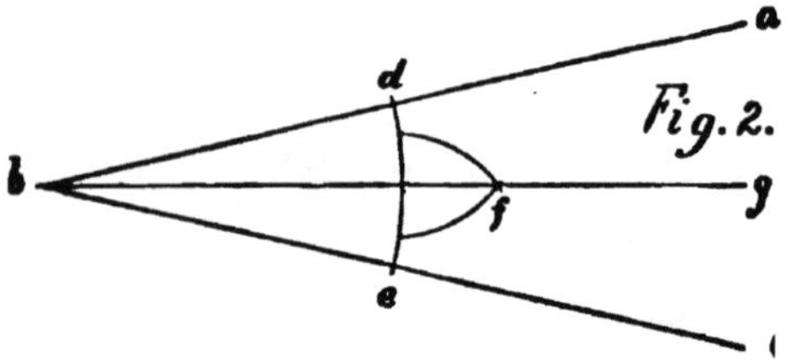

Fig. 2.

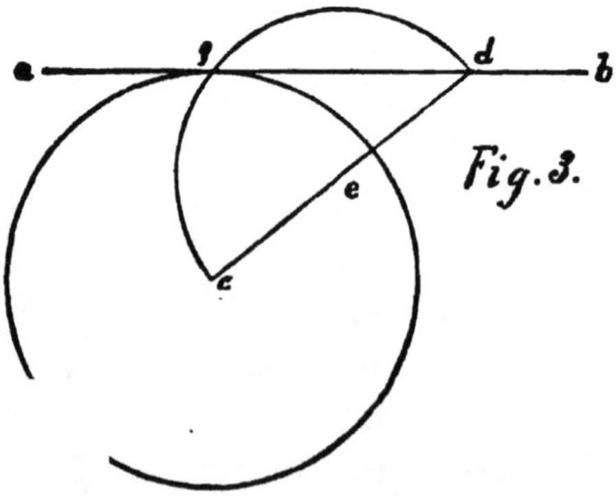

Fig. 3.

chalk-line. It would also be very convenient to use in describing a circle of great radius.

Plate 2. Fig. 5. *To describe an ellipse with the compasses.*—Draw the line *a b*, which represents the length of the required ellipse. Bisect this line (see Plate 1, Fig. 1), which gives *c e*. Make the length of *c d* equal to half the width of the required ellipse. Divide *c d* into three equal parts, the points of division being at *f* and *g*. Measure off from *a*, and also from *b*, the length of two of these parts; which gives the points *i* and *i* 2. Join *i* and *g*. Bisect this line, continuing the bisecting line until it intersects with the line *c e*. At *e* is the centre from which to describe the side of the ellipse, and the points *i* and *i* 2 are the centres from which to describe the ends. A line drawn from *e*, through the points *i* and *i* 2, will show where the curve of the sides and the curve of the ends meet, as seen at *j* and *k*.

Plate 2. Fig. 6. *To describe an ellipse with a two-foot square.*—Draw a line *a b* in the direction of the length of the required ellipse. Lay the square on the line so that the inside edge of the blade will be on the line, and the inside corner *e* will be in the centre of the ellipse. Then, with any strip of board, form a trammel as follows: an inch or so from one end drive a brad through at *f*, letting it project through about an eighth of an inch. From this point, measure off one-half the width of the ellipse. At this point, bore a small hole, and insert a piece of pencil, *g*, which must project down far enough to mark when the trammel is laid down on the square. Then, from this point, measure off one-half the length of the ellipse, and drive through a brad, *h*, letting it project below the same, as at *f*. Then, by sliding down on *h*, and letting *f* move to the left, all the while keeping *h* and *f* hard up against the edge of the square, the pencil *g* will describe one-quarter of an ellipse. Then turn the square over so that the end *a* will be in the direction of *b*, keeping the inside corner of the square on the point *e*, and describe the other quarter in the same manner, thus forming half of an ellipse, the other half of which may be described in the same manner, by reversing the end *i*.

This rule applies when the sum of half the length and half the width of the ellipse does not exceed the length of the tongue of the square. For larger ellipses, two straight-edged pieces of board might be used, one being *a e* and the other *e i*, which could be fastened to the work at right angles with each other.

Plate 3. Fig. 7. *To describe an elliptic arch by finding points through which to spring a lath.* — Let *a b* be the span or chord of the required arch,

Plate 2.

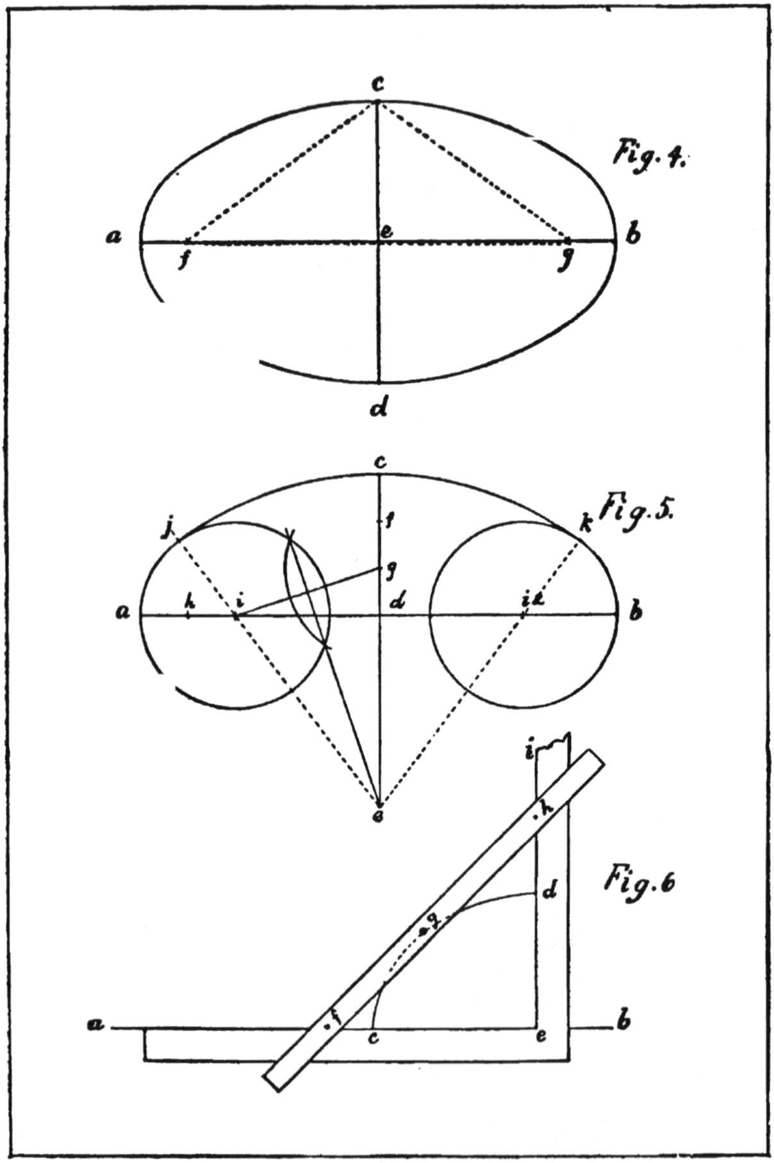

and *c d* be the rise. At *a* and at *b*, draw perpendicular lines, *a e* and *b f*, to the height of *c d*. Also draw a line joining *e* and *f*. Divide *d f* and *f b* each into any number of equal parts, as 1, 2, 3, 4, and 5, 6, 7, 8. Draw lines joining *d* and 5, 1 and 6, 2 and 7, 3 and 8, and 4 and *b*. Then through the points of intersection, *d*, *g*, *h*, *i*, *j*, and *b*, spring a thin strip of board, and mark around it. Repeat the operation on the other side.

This method is very much used by builders, but we prefer the method described in Plate 2, Fig. 4.

Plate 3. Fig. 8. *To describe a curve of great radius.* It is sometimes desired to describe a curve of great radius. The usual method is to use a line (for a radius to strike the curve), but a line stretches so as to give an irregular curve; and then, again, there is not always room to use a sufficiently long radius. The method described in Fig. 8 is the best way in such cases, when the rise and span are known, it being very quickly done, and giving a true curve.

Let *a b* be the span, and *c* be the rise. Tack in a nail at *a* and at *b*. Take two strips of board four or five inches wide, and six or eight inches longer than the span of the required curve. Joint straight one edge of each. Lay one piece with the straight edge in against *a c*, and lay the other piece with the straight edge in against *b c*, letting the ends lap at *c*, and drive in a nail. Also fasten a stay across, so that when the sides are against the nails at *a* and *b*, the corner of the frame will be at *c*. Then, keeping this frame against the nails at *a* and *b*, slide the frame around, holding a pencil at *c*: the pencil will describe a true curve, as shown in the dotted line. A piece of one-eighth inch round wire, two or three inches long, would be better than nails to tack in at *a* and *b*.

Plate 3. Fig. 9. *Given a segment of a circle, to find the centre.* — Mark off any three points on the segment, as *a*, *b*, *c*. With *a* and also with *b* for a centre, and a radius somewhat more than half of *a c*, describe the arcs *d e* and *f g*. Then, with *c* for a centre, describe arcs intersecting these, as shown in the cut. Through the points of intersection at *d e* and *f g*, draw lines, continuing them until they intersect at *h*, which point is the centre from which the segment *a b* was described.

When the segment is very large, or when it is desired to be very exact, the quickest and best way is to figure out the centre, which may be done as follows: Square half of the span; to this add the square of the rise, which sum divide by the rise: the quotient is the diameter of the circle, of which the given segment is a part. Thus, suppose the span *a b* is 60 inches, the rise, 10 inches:

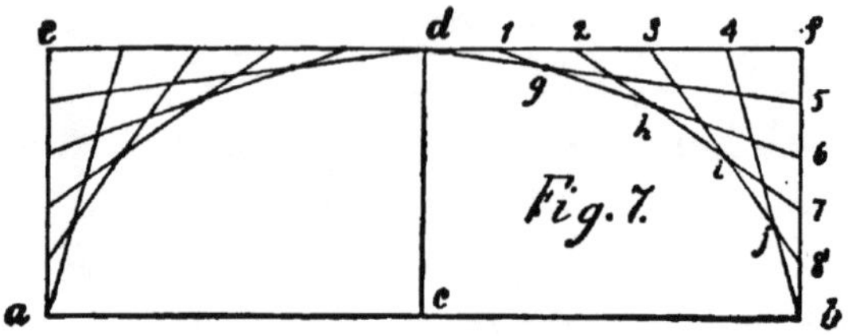

Fig. 7.

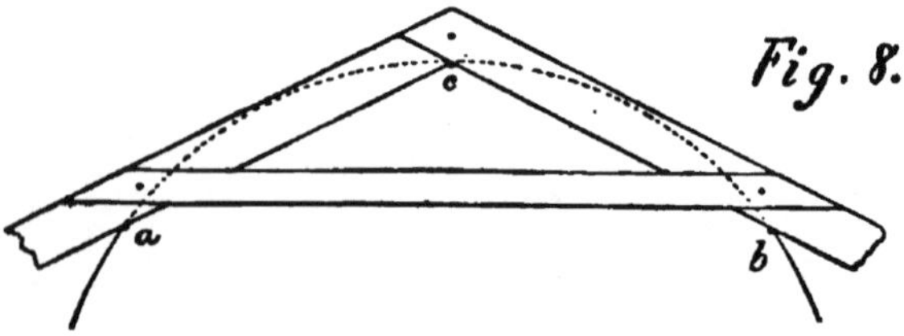

Fig. 8.

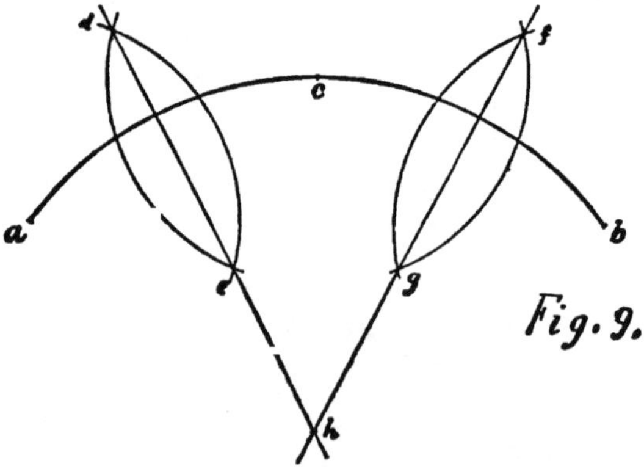

Fig. 9.

then half of the span is 30 inches, the square of which is 30 × 30=900. The square of the rise, 10 inches, is 10 × 10=100, which, added to 900, the square of half the span, makes 1,000; which, divided by the rise, 10, gives 100 inches — that is, 8 feet, 4 inches — as the diameter.[1]

Plate 4. Fig. 10. *To find how far apart to saw kerfs to spring a board or moulding.* — Let *a b* be the curve, around which it is desired to spring a piece of stock. Take a piece of stock *d g* of the thickness which is to be used; lay it down so that the edge shall pass through the centre *c*, and mark from *c* to *g*, and also at *e*. Now, with the saw which is to be used, make a kerf nearly through the piece of stock at *c*. Now, keeping this piece on the line *c g*, spring down the end *d* until the kerf is closed, then mark the point *f*: *e f* will be the distance apart to saw kerfs.

Plate 4. Fig. 11. *To describe a spiral.* — Draw a line *a b*, on which, near the centre, locate two points, *d* and *e*, which must be placed just half as far apart as it is desired to have the lines of the spiral. Midway between these two points is *c*, the centre of the circle from which the spiral begins.

Place one point of the compasses in *e*, and with a radius of *e* 1, describe the semicircle 1, 2.

[1] When the diameter is given, to find the rise for any chord or span, see p. 120.

Then, using *d* for a centre, and with a radius of *d* 2, describe the semicircle 2, 3. Then, again, with *e* for a centre, and with a radius of *e* 3, describe the semicircle 3, 4, and so on.

This rule does not give a true spiral, although it answers in most cases. To describe a perfect spiral, turn out a piece of wood, an inch long, of such size that the circumference of this piece shall be equal to the length of space between the lines of the spiral; that is, the diameter of this piece shall be about one-third of the distance between the lines of the spiral.

Fasten this turned piece in the centre of the intended spiral, and fasten one end of a piece of thread to this piece. Wind the thread around this piece, and make a loop in the last end of the thread. Now, holding a pencil plumb in this loop, swing the pencil around so as to unwind the thread, letting the pencil mark as the thread unwinds. The pencil will describe a true spiral

Plate 4. Fig. 12. *Given one side to construct an equilateral triangle.*—Let *a b* be the given side. First with *a*, and then with *b*, for centres, and with a radius equal to *a b*, describe arcs intersecting at *c*. Join *a c* and *b c*, which forms the required triangle. The arcs thus described, which are shown in dotted lines, also form a Gothic arch.

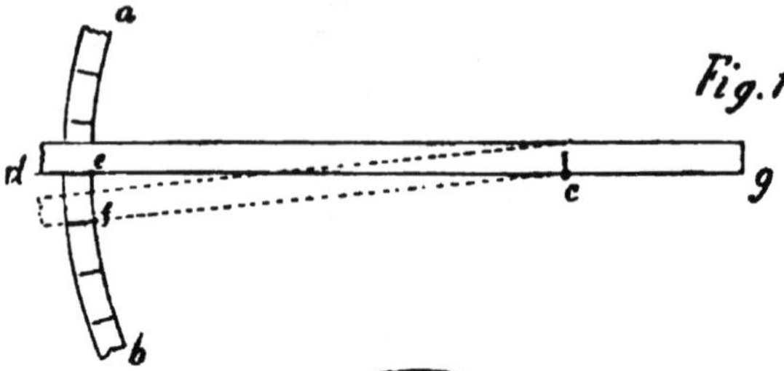

Fig. 10.

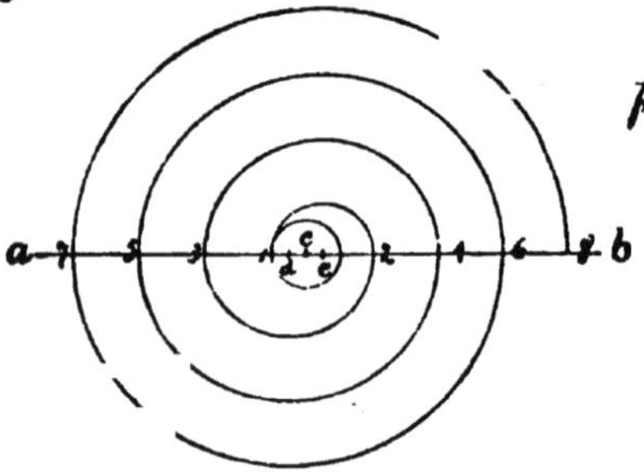

Fig. 11.

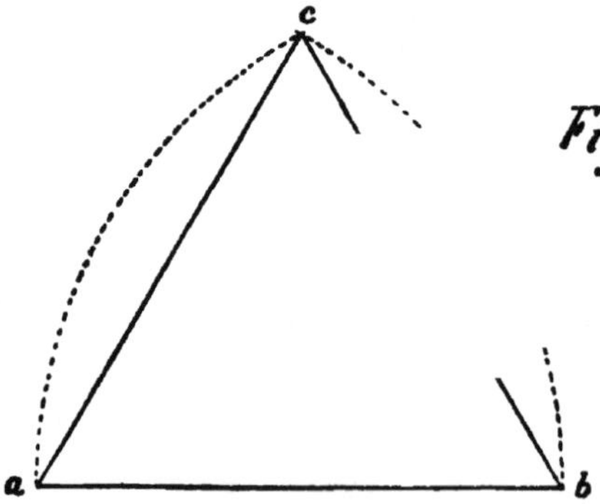

Fig. 12.

Plate 5. Fig. 13. *Given one side to construct a polygon of any number of sides.* — Let *a b* represent one side of a five-sided polygon (pentagon). Continue *a b* indefinitely toward *c.* With *a* for a centre, and a radius equal to *a b*, describe a semicircle, which divide into as many parts as there are sides in the required polygon. From *a* draw a line to the division 2. With these three points, *a*, *b*, and 2, find the centre of a circle, the circumference of which will pass through them. (See Plate 3, Fig. 9.) Then space off this circle, making the spaces the length of the given side *a b*, which gives the points *a*, *b*, *e*, *d*, and 2. Join these points, and we have the required polygon.

Plate 5. Fig. 14. *Given the distance across, to construct a six or eight sided polygon (hexagon or octagon).* — Draw a circle, the diameter of which equals the distance across the required polygon, through the centre of which draw the line A B. Space the circle into six or eight parts, as may be required, and draw lines from the centre through the points of division. Join *f* and *d*. With a pair of compasses take the distance which the centre, *b*, of this line falls short of the point *a*, and lay off the same from *f* to *e*; also from *d* to *c*. Now draw a line joining *c* and *e*, continuing the line to the point A. Then with *g* for a centre, and a radius of *g* A, describe a circle. Join the points where the radial lines cross this circle; and the result is the required polygon.

This method is used when part of a piece of turned work is to be six or eight squared, and the distance across the squares is given.

Plate 5. Fig. 15. *To construct an octagon from a square.*—Let *a*, *b*, *c*, *d*, be the given square. Join *a* and *d*. With *c* for a centre, and a radius tangent to this line, describe an arc intersecting the sides of the given square. Then, with the points *a*, *b*, and *d* for centres, and with the same radius, describe other arcs in the same manner. Join the points of intersection as shown in the cut. The result is the required octagon.

Plate 6. Fig. 16. *On a given diagonal to construct a square.*—Let *a b* be the given diagonal. Bisect *a b* (see Plate 1, Fig. 1), getting the line *c d*. Take the point of intersection, *e*, for a centre, and with a radius equal to *a e*, cut *c* and *d*. Join *a d*, *c b*, *a c*, and *d b*, and we have the required square.

Plate 6. Fig. 17. *To inscribe a circle in a triangle; also to describe a circle around a triangle.*—To inscribe a circle in a triangle, bisect any two of the angles (see Plate 1, Fig. 2), continuing the bisecting lines until they meet, which point is the centre from which to strike an inscribed circle.

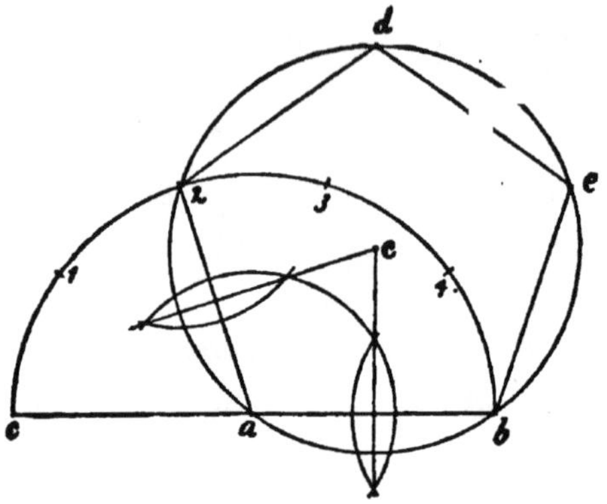

Fig. 13.

Fig. 14.

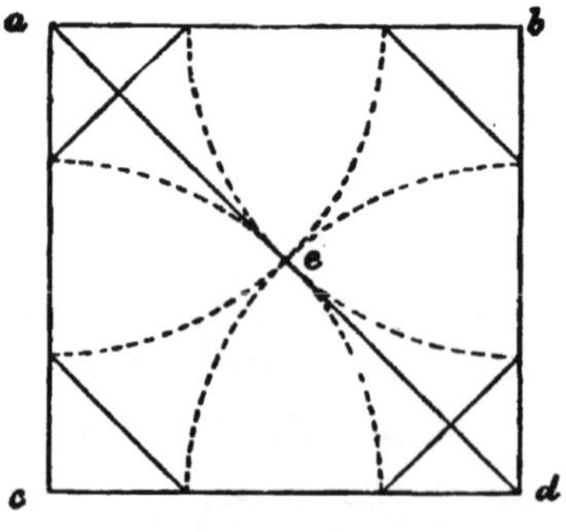

Fig. 15.

To describe a circle around a triangle, bisect any two of the sides, continuing the bisecting lines until they meet, which point is the centre from which to describe a circle around the triangle.

Plate 6. Fig 18. *To inscribe an equilateral triangle in a circle.*—Through the centre of the circle, draw the vertical line *a b*. With *b* for a centre, and a radius the same as used to describe the given circle, describe an arc intersecting at *d* and *e*. Join *a d*, *a e*, and *d e*, which forms an inscribed triangle.

Plate 7. Fig. 19. *To inscribe a square in a circle.*—Through the centre of the circle, draw the line *a b* at an angle of 45 degrees. (See Plate 8, Fig. 22.) Bisect this line, producing the line *c d*. Join *a d*, *c b*, *a c*, and *d b*, which forms the inscribed square.

Plate 7. Fig. 20. *To inscribe a hexagon in a circle.*—The radius used to describe the circle will space around the circumference just six times; and, by joining these points, the required inscribed hexagon is formed.

Plate 7. Fig. 21. *To describe the envelope of a cone.*—Let A be the apex of the cone, and B C be the base. The rule commonly given is, to use

A for a centre, and with a radius of A C describe an arc, as shown at C D. Then, with the radius used to describe the plan of the base,—the diameter of which is B C,—lay off six spaces. (The cut, for lack of room, shows only half of them.) Draw lines joining the first and the last of these spaces to A.

This is not exact. Lay off on each side of the cone the thickness of the envelope, which gives *a b c*, which is to be considered as the cone. Then, with *a* for a centre, and a radius of *a c*, describe an arc the same as previously described. Then find the circumference of the base of the cone, the diameter of which is *b c*. This is found by multiplying the diameter *b c* by 3.1416, or $3\frac{1}{7}$. This length is to be measured around the curve of the base of the envelope, which determines the length of the envelope. Then join these ends to *a*, which gives the form of the envelope.

If the cone is truncated, that is, the top cut off, as shown at *x y*, then *y z* shows the top of the envelope.

Plate 8. Fig. 22. We give here a scale of degrees, commonly called a Protractor, which we believe will be found quite convenient. The various mitres and angles may be taken from this protractor by placing a bevel with the stock on the line, as shown in the cut, and running the tongue from the

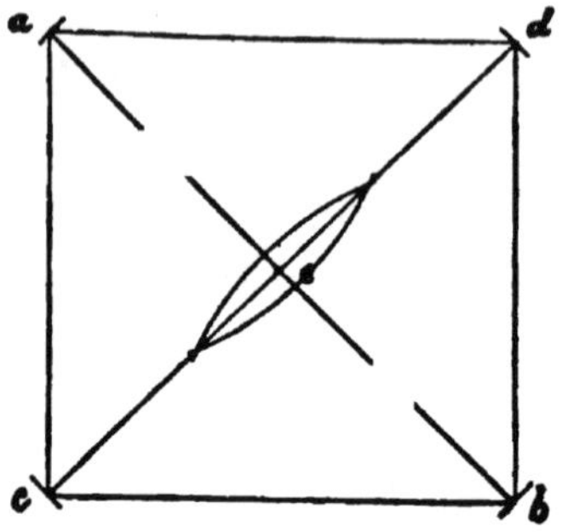

Fig. 16.

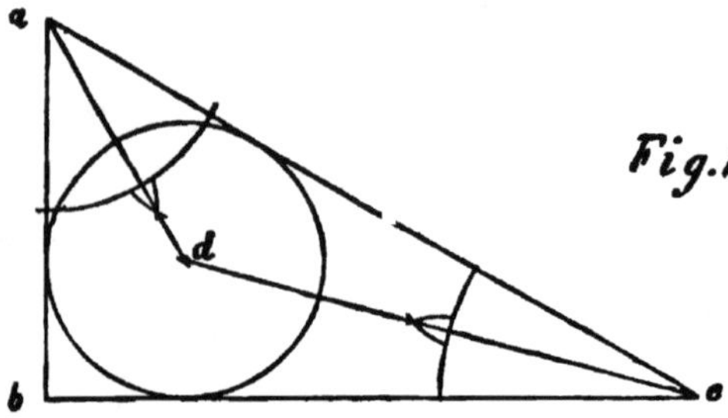

Fig. 17.

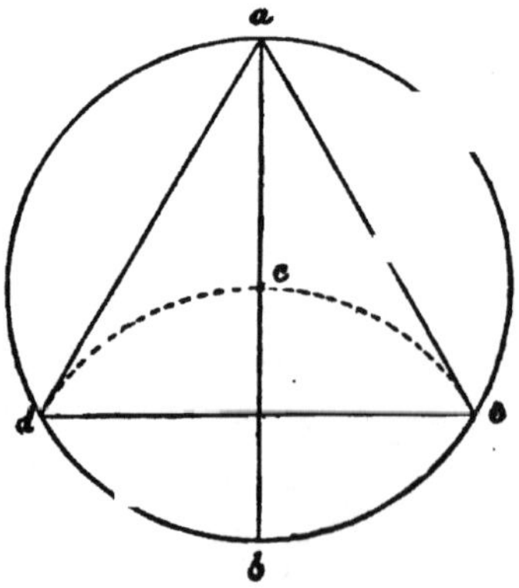

Fig. 18.

point *a* to the number of degrees required. The degrees continue on from 90 to 180; although we have not divided or numbered them, as we have those from 0 up to 90.

The angles of an equilateral triangle are 60°; the mitre is 30°.
The angles of a square figure are 90°; the mitre is 45°.
The angles of a hexagon are 120°; the mitre is 60°.
The angles of an octagon are 135°; the mitre is 67½°.

The sills of window-frames are usually set at an angle of 10 degrees. A few builders set them steeper. We have seen some set at an angle of nearly 20 degrees.

Plate 9. Figs. 23, 24, 25, and 26, illustrate different methods of splicing timbers. Fig. 26 is a keyed diagonal splice, the shoulders being cut square with the slant of the splice. The shaded part is a hard-wood key.

Fig. 27 is bridging for floors. Common strapping 1×3 inches is generally used; although for heavy floors 1¼×3 or 4 inches may be used, being fastened with two good nails at each end.

Fig. 28 represents two timbers tied together, and supported by a braced post. The notching in the post and timbers for the braces should be cut square with the slant of the braces, as shown in the cut.

Plate 10. Fig. 29 *shows a plan for floorings.* The timbers are usually gained into the sills 2 inches, and down 4 inches, so as to bring the top of the timbers even with the top of the sills. The plan shows an opening for stairs. The headers *b b*, and the trimmer *a*, which is also shown in Fig. 30, are made of extra thickness: where the floorings are 2 inches thick, the headers and trimmer should be 3 inches thick.

The end sills should be 7 or 8 inches wide, so as to get a good nailing for the ends of the upper floor-boards, as shown in Fig. 31; while if the sills are narrow, as seen in Fig. 32, the ends of the upper floor-boards have no timber to nail into.

Plate 11. Fig. 33. *An end elevation of a two and one-half story dwelling-house.*—The dotted lines at *g g* show the position of the girts or ledger-boards on the side of the building, being put down so that the floorings may set on them, and come even with the top side of the end girt.

Plate 12. Fig. 34. *The side elevation of the same house as Fig. 33, being represented with side girts.*—Another way, and in some respects to be preferred to this way, is to use ledger-boards instead of girts, which allows the studding to run whole length from the sill to the plate. Braces may be put from the sills to the posts, and from the plates to the posts. With girts there are more chances to put braces.

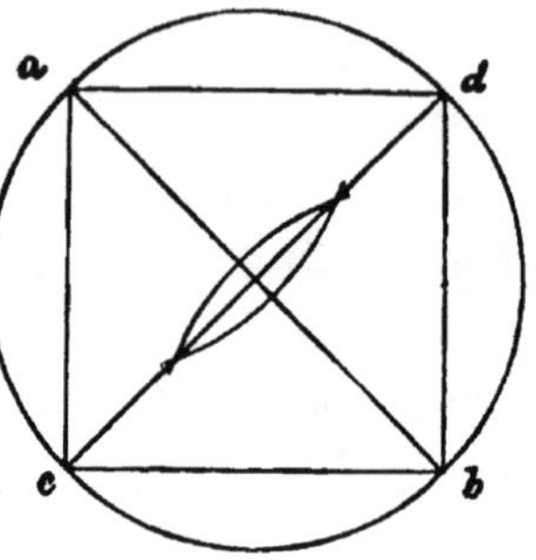

Fig.19.

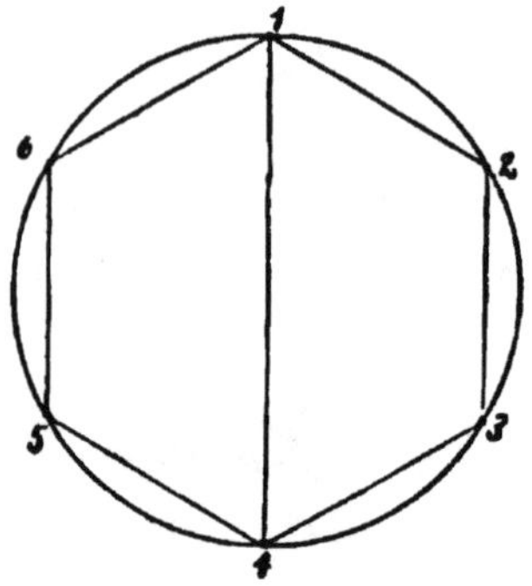

Fig.20.

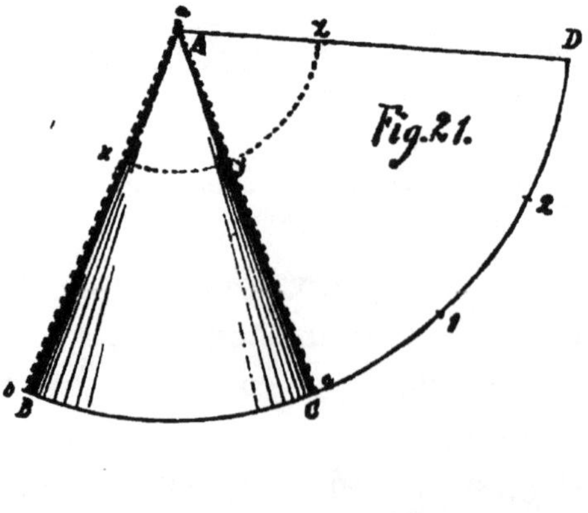

Fig.21.

Plate 13. Fig. 35 shows the method of finding the bevels of rafters for pitch roofs. Let *a b* be the width of the building, which may be drawn to the scale of one and one-half inches to the foot, — each one-eighth inch of the drawing representing one inch, — and *c d* be the rise. Join *a* and *c*, which gives the pitch of the roof. At *c* is seen the bevel for the top of the rafter, and at *a* is seen the bevel for the rafter where it rests on the plate as shown at *a e*, Fig. 36.

Fig. 36 shows the manner of laying out a rafter. The crowning edge of the timbers should always be the outside edge of the rafters. Having laid out and made off one rafter, use it as a pattern with which to lay out the others, keeping them even at *b*, and at the top end *c*. When there is a ridge-piece, cut off from the end of the rafter half of the thickness of the ridge-piece, measured square from *c d*. (See also Plate 14, Fig. 38.)

Fig. 37 shows a way of getting the length and bevels of rafters with a two-foot square. Have the outside edge of the rafter next to you. Suppose that the width of the building is 20 feet, and the rise of the roof is 7 feet. Let inches on the square represent feet on the building. Take half the width of the building — 10 inches — on the blade of the square, and take the rise — 7 inches — on the tongue. Hold the square as shown in the cut, having these points even with the top edge of the rafter. The bevel on the rafter at the blade of the square is the bevel of the rafter where it sets on the plate as seen at *a e*, Fig. 36. The bevel on the rafter at the tongue of the square is the down bevel for the top of the rafter. Now, as the measures on the square were in inches, while those on the building were in feet, it follows that the diagonal from 10 inches on the blade to 7 inches on the tongue of the square is $\frac{1}{12}$ of the length of the rafter: so, by measuring off 12 times this length, we have the length of the rafter. Where there is a ridge-piece, do as directed in Fig. 36.[1]

Plate 14. Fig. 38 represents the rafters of a pitch roof. Fig. 39 represents the rafters of a hip roof. If the rafters on a pitch roof are 2 × 6 inches, they should be notched for the plate so as to leave the rafter 4 or 4½ inches at the narrowest point; then measure the perpendicular width at this point, as indicated by the line A *a*. Subtract this amount from the rise of the roof, and it gives the rise to use in getting the bevels for the rafters as described in Fig. 35, Plate 13.

In framing the rafters for hip roofs, Fig. 39, there is not usually so much stock in the rafter above the plates as there is in rafters for pitch roofs; the lower end of the rafter being dropped in order to have sufficient stock to form a crow-foot.

[1] The lengths and bevels of braces may be found in a similar manner. Suppose the run is 36 inches by 48 inches, we may take any fractional part of the run on the square, say, for instance, one-third, which will be 12 inches on the tongue, and 16 inches on the blade of the square: then three times the diagonal

Plate 8.

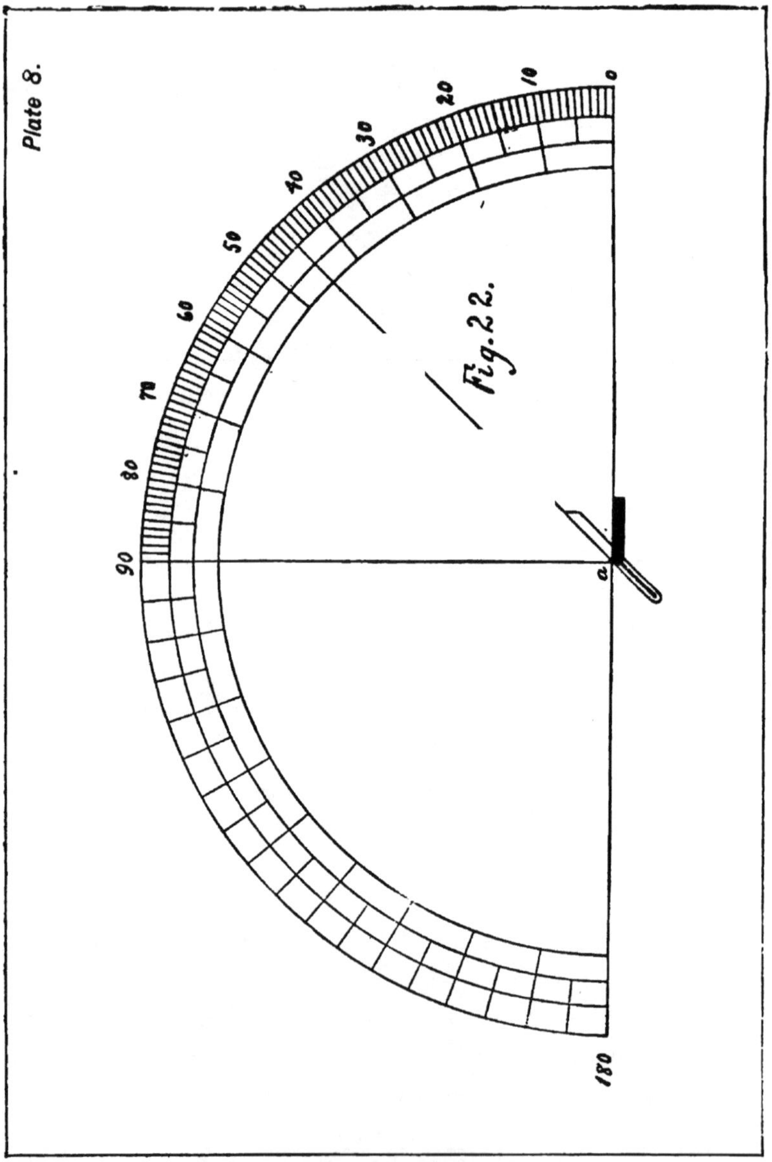

Fig. 22.

Fig. 40 shows the three pitches in common use. The pitch at *e* is called the square pitch, the slant of one side of the roof being at right angles to the slant of the other side, the pitch of the roof being 45 degrees. The pitch at *d* is $\frac{2}{3}$ pitch, the length of the rafters being $\frac{2}{3}$ of the width of the building. The pitch at *c* is called $\frac{1}{3}$ pitch, the rise being $\frac{1}{3}$ of the width of the building. There is also the Gothic pitch where the length of the rafters is equal to the width of the building.

Plate 15 shows the method of getting the lengths and finding the bevels of rafters for hip roofs. Fig. 41 is the elevation of the roof, *a b* being the width of the building, and *c d* being the rise of the roof; *a d* and *b d* are the length of the common rafters, the bevels of which are found in the same manner as the bevels of rafters for pitch roofs.

Fig. 42.—*a b c d* is the plan of the building; *e f* is the plan of the ridge-piece; *a f*, *b f*, *c e and d e*, is the plan of the hip rafters. Draw the line *g h*, the length of the common rafter *a d*, square with the line *a c;* and, passing through *e*, join *c* and *h*, which gives the length of the hip rafter; draw the line *o p* through *h* parallel to *e f;* the edge bevel of the hip rafter is shown at *h** and the edge bevel of the jack rafters is shown at *j;* the lengths of the jack rafters are *m n*, *k l*, and *i j*, the down

* ***First*** **back off the upper edge of the hip rafter,** ***then*** **use this bevel. The rafter will** ***not*** **fit if this bevel is used before the rafter is backed off.**

bevels being the same as the down bevels of the common rafters.

To find the down bevels for the hip rafter, Plate 16, Fig. 43, make *a h* equal to the length of the plan of the hip rafter (*a f*, Fig. 42), and make *g h* equal to the rise of the roof, *c d;* join *a* and *g*, which gives the elevation of the hip rafter; the bevel for the foot being shown at *a*, and the down bevel for the top being shown at *g*.

To find the backing of the hip rafter (that is, the amount necessary to chamfer the top edge), take any point on the line *a b*, Fig. 41, as *e;* draw a line through this point, square with the slant of the roof, as seen at *e f*. Take the distance from *a* to *e* (Fig. 41), and lay it off from *a* to *r*, and from *a* to *s* (Fig. 42). Join *r* and *s*. Take the distance from *e* to *f* (Fig. 41), and set it off from *t* to *u* (Fig. 42). Join *r u* and *s u;* then *r u s* is a section of the roof cut across the corner at *r s*, and cut down square with the slant of the roof, as seen at *e f* (Fig. 41). Now lay off the thickness of the hip rafter, equally each side of the line *t u*, and we have the shape of a section of the hip rafter; and the amount necessary to chamfer may be seen, or a bevel may be set at the angle formed by the lines *u r* and *u t*.

In getting these bevels, the work may be drawn to the scale of 1½ inches to the foot, each ⅛ inch representing one inch of the work.

Fig. 23. Fig. 24.

Fig. 25. Fig. 26.

Fig. 27.

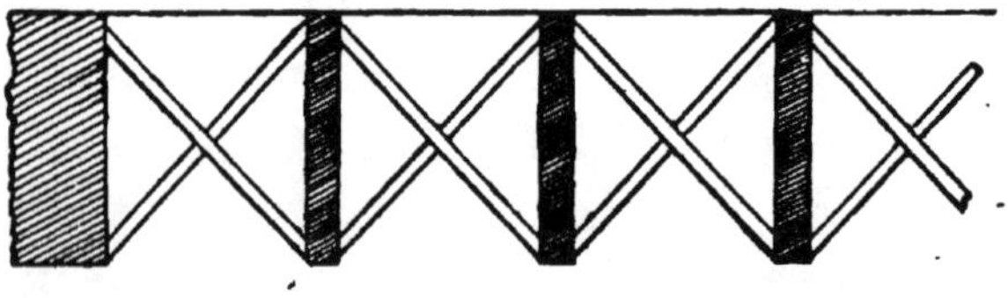

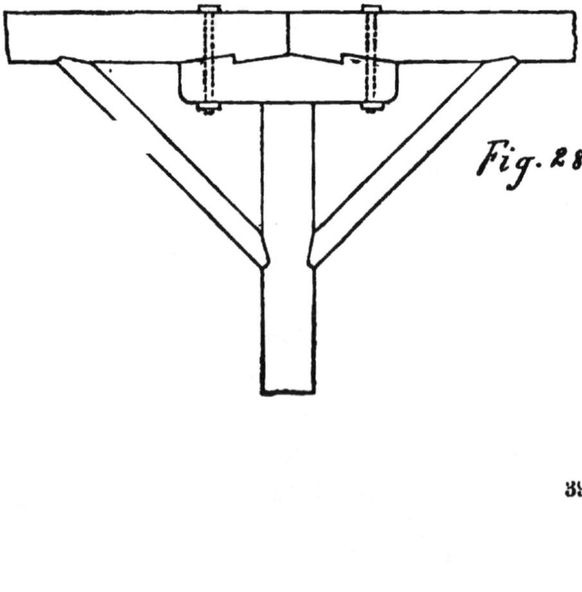

Fig. 28.

If the hip roof is placed above a French or mansard roof which tumbles in, then the drawing of the plan of the hip roof must be made of the size of the upper plates, instead of the size of the building.

Greater accuracy is obtained by figuring out the lengths of the hip and common rafters. To find the length of the common rafters, square half of the width of the building; also square the rise of the roof; add together these two amounts, and extract the square root (see the mathematical part of this book): the result is the length of the common rafter. To find the length of the hip rafter, square the length of the common rafter; also square half the width of the building; add together these two amounts, and extract the square root: the result is the length of the hip rafter, on the centre line, as seen at *a h*, Fig. 44. From these lengths half of the thickness of the ridge-piece must be deducted in the manner before described. The dimensions should be taken in inches. This rule to figure out the lengths applies only to square buildings, and where the pitch of the roof is the same on all sides. Where the buildings are not square, then draw a plan of the whole roof, and find the lengths and bevels of the hip and jack rafters for each of the different corners, as described in Fig. 42.

Plate 16. Fig. 44 shows the framing of a hip roof. The centre lines in the hip rafters and in the ridge-piece are the lines representing the plan of the hip rafters and the ridge-piece in Fig. 42, Plate 15. For convenience of fastening the hip rafters to the ridge-piece, this piece is made two or three inches longer at each end than *h j*, and the rafters *n n* are cut as much short of their whole length as the ridge-piece extends beyond *h* or *j;* the ridge-piece being scarfed from the points *h* and *j*, to the pitch of the rafters, *n n*.

Plate 17 shows the method of getting the lengths and finding the bevels of rafters for valley roofs. Fig. 45 is an elevation of the roof; *a b* is the width of the building, *c d* is the rise of the roof (see Plate 14, Fig. 38, deduct *a* A from the whole rise, and then consider C D as the rise); *b e* is the rise of the cross roof, and *f e* is the ridge of the same. The bevel for the foot of the common rafters of the main roof is seen at *b;* at *d* is seen the down bevel for the top of the common rafters, and for the top and bottom of the jack rafters, 4 5, and 6 *w*.

Fig. 46.—*a b c d* is a plan of the roof; *e f* is the ridge of the main roof; *g h* and *i h* is the plan of the long valley rafters; *j k* and *l m* is the plan of the short valley rafters; *z m* and *k p* is the plan of the ridge of the cross roof. Lay off from *d* to *n* the length of the common rafter *a d* (Fig. 45);

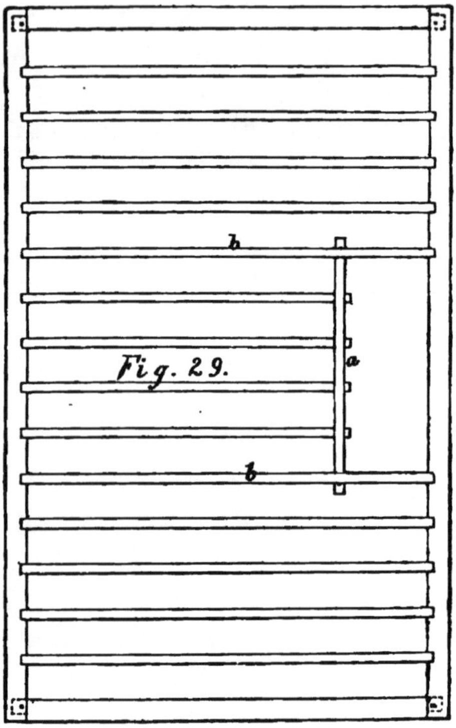

Fig. 29.

Fig. 30.

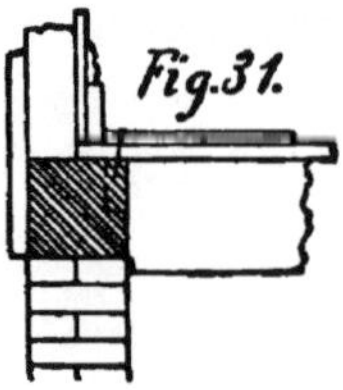
Fig. 31.

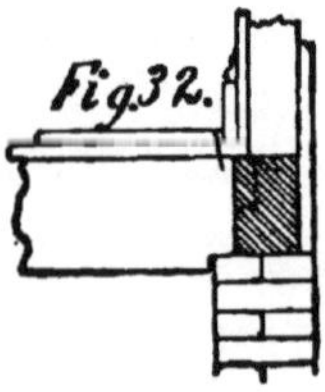
Fig. 32.

draw the line *n o* parallel to *e f;* then from *h*— the point where the plan of the valley rafter joins the ridge — draw the line *o h* square with the line *e f;* join *o* and *g*, which gives the length of the long valley rafter, the edge bevel for the top being shown at *o*. Draw a line from *j* to *k* 1, the point where the line *o g* crosses the ridge *z p;* then *j k* 1 is the length of the short valley rafter, the edge bevel for the same, where it butts against the long valley rafter, being shown at *k* 1. The length of the jack rafters is seen at 4 5, and 6 *w*, the edge bevel for the same being shown at *w*. To find the down bevels of the valley rafter, take the plan of the valley rafter *i h* for a base line; from *h* draw the line *x h;* square from *i h*, making *x h* equal to the rise of the roof *c d* (Fig. 45); join *i* and *x*, which gives the elevation of the valley rafters, the bevel for the foot being shown at *i*, and the down bevel for the top being shown at *x*. To find the length and down bevels of the rafters for the cross roof, take *g p* for a base line; draw the line *p r* at right angles with *p g*, equal to the rise of the cross roof *e b* (Fig. 45); join *r* and *g*, which gives the length of the common rafters (when there are any, as when the building is made in the form of a cross, +); the bevel for the foot of the common rafters being shown at *g*, the down bevel for the top of the common rafters, and the top and bottom of the jack rafters, being shown at *r*. Lay the length of the common rafter *g r* from *g* to *s;* draw *s t* parallel to *p k;* draw the line *t k* from *k*, square with the line *k p;* then draw a line representing the valley rafter from *g* through the point *t;* then the length of the jack rafters is shown at 1, 2, and 3 *v*, the edge bevels for the same being shown at *v*. The lengths of these rafters are measured on the centre line of the edge; and from these lengths, half of the thickness of the ridge-piece, or half the thickness of the valley rafter, must be cut off, as may be necessary; the amount to be cut off must be measured square from the bevel on the end of the rafters, so that if the ridge-piece was two inches thick, the piece which would be cut off would be one inch in thickness.

Sometimes the ridge of the cross roof is carried clear through from *z* to *p;* in this case the valley rafters are all of the same length. The plan of them would be *g k*, *j k*, *i m*, and *l m*, their actual length being *g t;* and the edge bevel of the top of the valley rafters is the angle formed by the lines *g t* and *s t*. *d g n o* is part of the roof laid down flat, the edge being kept even with *d g;* now, if *n o* is raised equal to *c d* (Fig. 45), it will stand just plumb over *f h;* and the hip rafter *o g* will stand plumb over the plain of the valley *h g*. *g t s* is one side of the cross roof laid down flat, the foot of the rafter *s g* being kept even at *g;* now, if *t s* is raised the height of *e b* (Fig. 45), the end *s* of the rafter

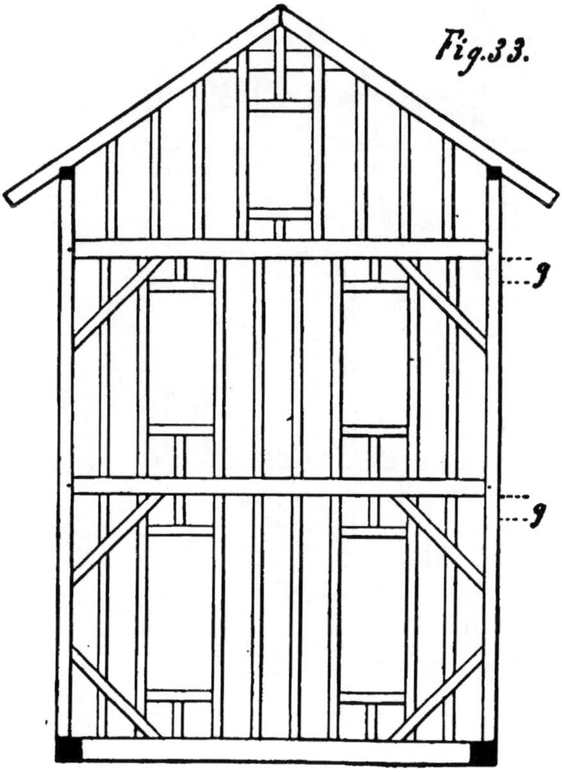

Fig. 33.

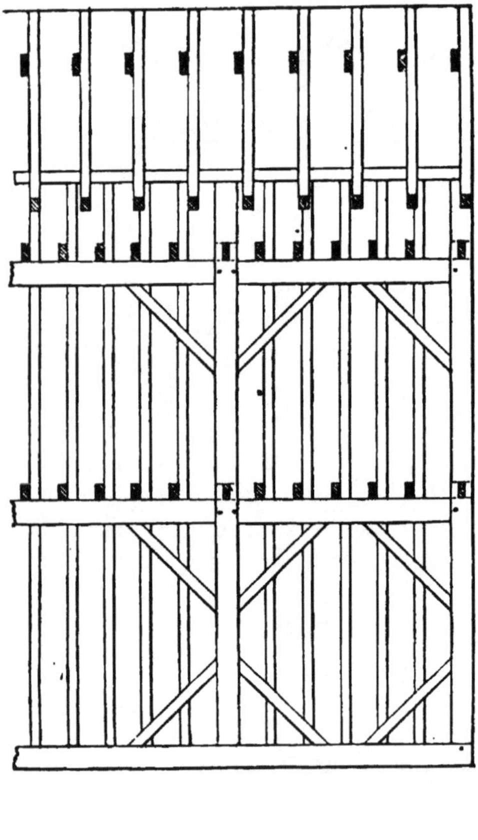

Fig. 34.

g s being kept plumb over the plate *d b*, then *s t* will stand plumb over *k p*, and the lines *o g* and *u g* will meet, and stand plumb over *h g;* the points *t* and *k* 1 will meet plumb over *k*.

The line of the top side of the jack rafters must run to the centre of the edge of the valley rafters; and bevelled strips may be nailed along on each corner of the upper edge of the valley rafter, so as to continue the slant of the roof from the end of the jack rafters to the centre of the edge of the valley rafters, as shown at *y*. In laying out such work as this, it is well to draw it to as large a scale as possible, using a hard pencil, sharpened fine, and work accurately: since, if your drawing is $\frac{1}{10}$ size, a variation in the drawing of $\frac{1}{16}$ of an inch will cause a variation of $\frac{5}{8}$ of an inch in the work; errors in the drawing being magnified ten times in the work.

Plate 18. Fig. 47 shows a frame of a mansard roof; *a* being the studding of the house, *b* the ledger boards, *c* the floorings, *d* the plate of the house, *e* the mansard rafter, *f* the roof-plates, *g* the rafter of the hip roof; *h* represents pieces of plank fastened to the studding, being boarded across and tinned on top, the gutter being nailed to the outside ends, thus forming a coving. A French roof is usually formed by nailing sweeps, *i*, which are made of plank, to the straight rafters *e*. Fig. 48

shows another method of forming a French roof. By this method the outside studding runs to the hip roof, the sweeps for the French roof being fastened to these studs. This method gives rooms in the roof of the same size as those in the story below; and, if properly proportioned, it makes a very good-looking roof.

Plate 19. Fig. 49. *To describe the corner rafter on a French roof.* — Let L *c m* represent the plan of a corner of the building, which in this case is square. Bisect the angle L *c m* (see Plate 1, Fig. 2), which gives the line *g c*, which is the centre line of the plan of the rafter, on each side of which lay off half the thickness of the corner rafter. Draw the line *c b* square with the line L *c*. Measure off from the line L *c*, the perpendicular length (not the length on the slant) of the straight rafter *e*, and draw the line *a b* square with *b c*. On this line *a b*, measure off from *b* to *f* the amount which the straight rafter tumbles in. Now place one of the common curved rafters against the line of the straight rafter *c f*; and, keeping the bottom of the rafter on the line *a b*, mark out the shape of the common curved rafter. Now draw the lines 1, 2, 3, 4, 5, 6, parallel to *b c* from various points of the curve, running them to the centre line of the plan of the angle rafter *g c*. Then draw these lines square from the line *g c*, making the length

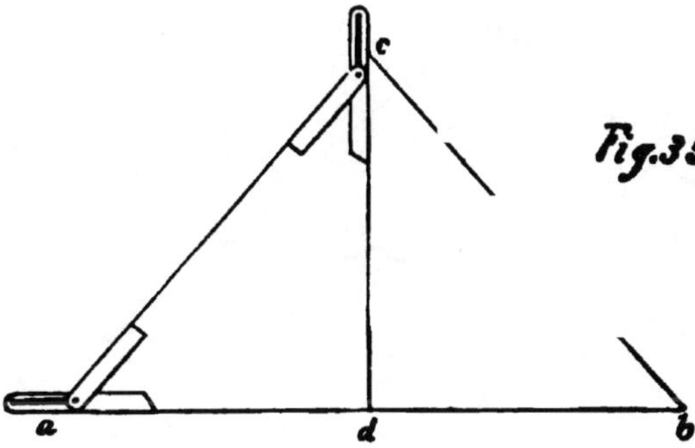

Fig. 35.

Fig. 36.

Fig. 37.

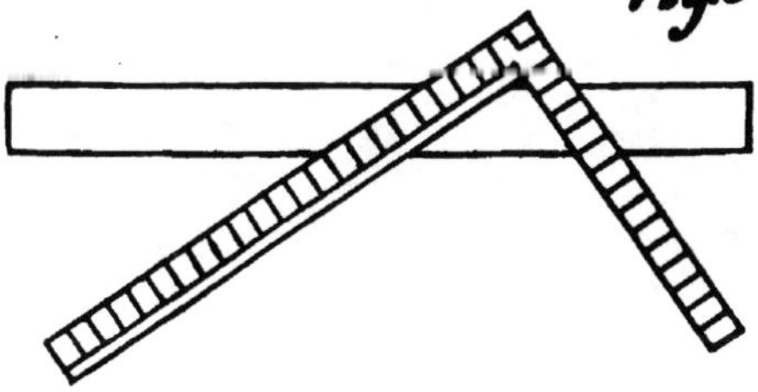

Plate 14.

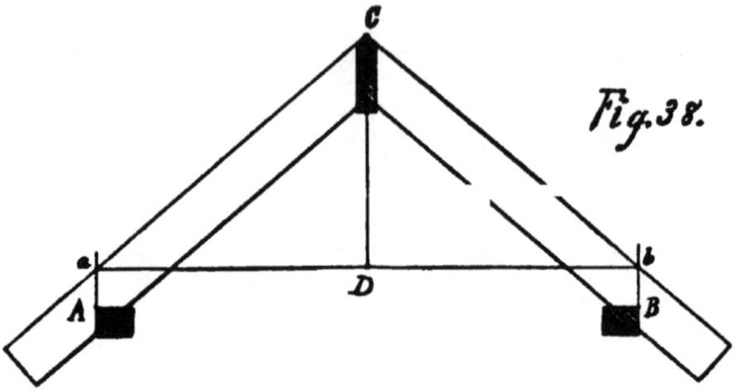

Fig. 38.

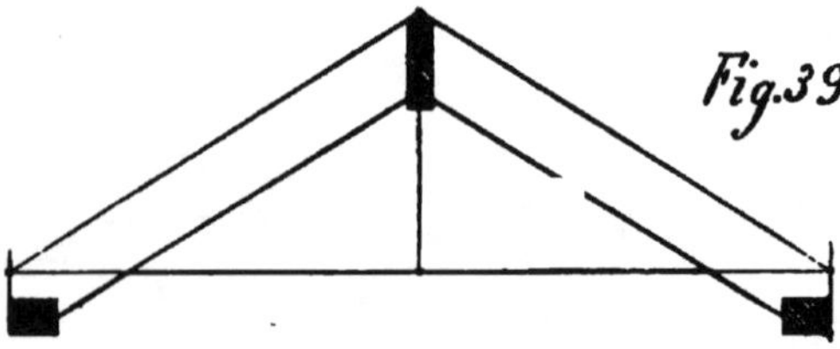

Fig. 39.

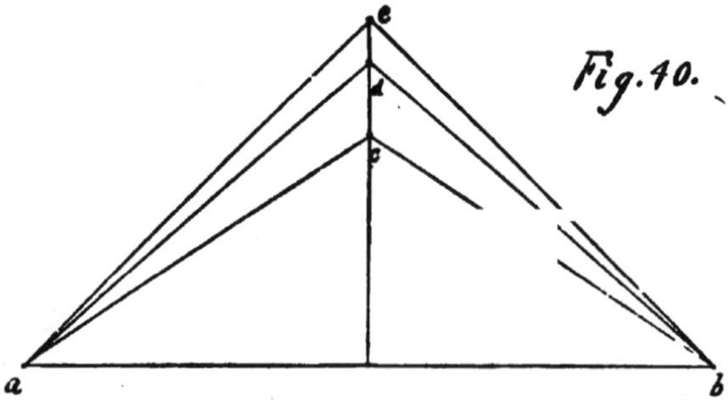

Fig. 40.

of the line 1 *a* to D the same as the line 1 from *d* to the line L *c;* and making the line 2 *a* the same length as the line 2 to the line L *c*, and so on. Draw a curved line from D through these points to E, which gives the curve of the edge of the angle rafter (from E to *c* will be straight). Make D A the length of *d a*. The line *c* B is drawn from the point *c*, square with the line *c g;* the line *c* B being drawn the length of *c b*. Join A and B, and we have the shape of the corner of the roof plumb down from the top of the straight rafter *c*. Line 5 runs to *f*, the foot of the straight rafter: so the line 5 *a* will determine the position of the foot F of the straight angle rafter. The line 6 runs from *e*, the top of the curved rafter, so the line 6 *a* will run to E, the top of the corner curved rafter. Then the shape of the corner curved rafter is E D A F. To find the splay, or chamfer, draw lines parallel to 1 *a*, 2 *a*, etc., from the point where the lines 1, 2, etc., pass through the line representing the edge of the rafter *g c*. Also draw lines square from the ends of 1 *a*, 2 *a*, etc., to these lines. Now draw a curved line from D through the points of intersection, and we have the amount necessary to chamfer the rafter. The length of the straight corner rafter is from *c* to F, which is somewhat more than the length of *c f:* and it is set with the centre of its edge exactly even with the corner of the building, as shown in Fig. 50 at *b;* 1, 2 being the sides of the rafter, *a b c* being a plan of the corner of the building, and R being the curved rafter. If it is desired not to chamfer the edge of the corner curved rafter, it must be sawed on the line of the chamber D *x*; and *y* would be the foot of the straight rafter. In this case, the straight rafter must be set with its corners even with the edge of the plate, as seen at 5 6, Fig. 50; that is, providing the thickness of the corner straight rafter is the same as the thickness of the corner curved rafter. Should the straight rafter be thicker, then gauge off the thickness of the curved rafter in the centre of the edge of the straight rafter, and let the lines representing the thickness of the curved rafter come even with the edge of the plate, as seen in Fig. 50; 3 4 7 8 represents a thick, straight rafter 5 6 being the thickness of the curved rafter, which points are set even with the edge of the plate. But if the corner curved rafter is chamfered, then the corner straight rafter must be set out even with the corner *b*, so that the points 5 6 would be in the place occupied by 1 2; and as much of the straight rafter as is above the curved rafter must be chamfered to correspond with the chamfer of the curved rafter. This rule applies for external and internal angles, whether they be acute, obtuse, or right angles. One thing, however, must be observed: that is, the rafter must be stayed from the building at the angle *c g*,

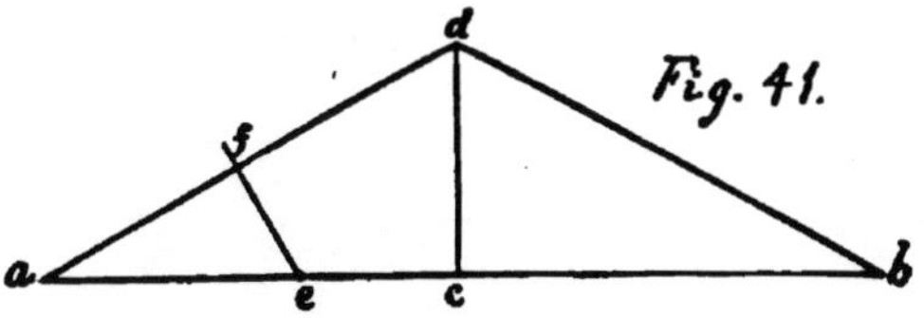

Fig. 41.

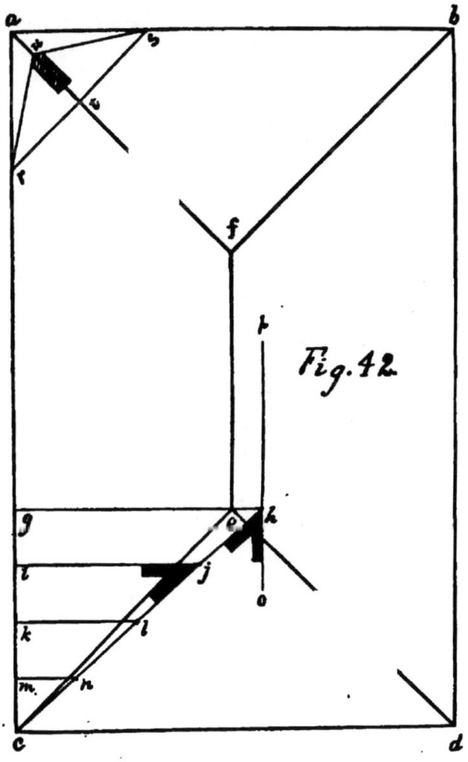

Fig. 42

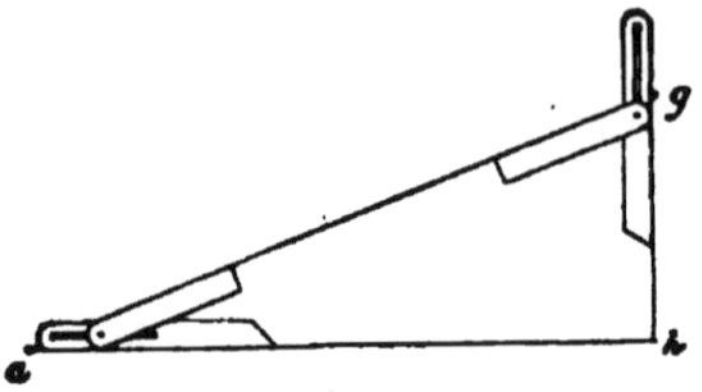

Fig. 43.

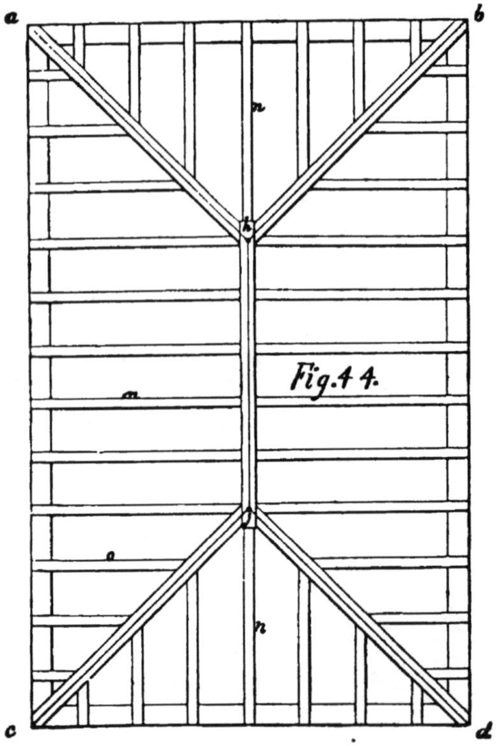

Fig. 44.

found by bisecting the angle formed by one corner of the building. Otherwise, it will not coincide with the line of the roof on both sides. This rule is also used for getting the angle brackets for large cornices, where they are lathed and plastered, and for getting the angle rafters for groined arches, etc.

Plate 20 represents three different forms of trusses. Fig. 51 represents the form of a truss suitable for a span of thirty or forty feet. Fig. 52 represents the form of a truss suitable for a span of forty or fifty feet. Fig. 53 represents the form of a truss suitable for a span of about seventy feet. These trusses are sometimes made with greater span than is here given.

Plate 21. Fig. 54 represents the form of truss suitable for short bridges, etc.; the bottom timber resting on abutments at A and B. The shaded pieces should be made of well-seasoned hard wood, or cast-iron. This form of truss is also used for supporting roofs of great span, as in halls and churches; the roof being supported from the truss by struts, etc.

Plate 22 represents the framing of a small spire; Fig. 55 being the elevation, and Fig. 56 the plan, of the spire. The rafters are sawed square on the edge at the top, and are fastened to the piece C by wooden pins, giving chance to bore down through C for a vane or finial. This is a much simpler way than to mitre the rafters together at the top. The backing (shown at *e*) is found as described in Plate 15, Fig. 42.

Plate 23 shows the method of finding the forms of the boards for boarding a dome roof horizontally. Fig. 57 is a plan of the boarding of the dome, and Fig. 58 is the elevation of the same. As will be seen in Fig. 58, the principle is the same as finding the envelopes of truncated cones. (See Plate 7, Fig. 21.)

Plate 24 shows the method of finding the form of the boards for boarding a dome roof vertically. Fig. 59 shows the elevation of the dome; and Fig. 60 is the plan of the same, the circumference of which we divide into spaces equal to the width of the boards to be used. *a b* C (Fig. 60) is the plan of one of these boards. The length of one of these boards is B C, Fig. 59, which we divide into any number of equal parts. Then from *a b*, Fig. 60, lay off the same number of these spaces to *c*. Then, from these points of division in Fig. 59, drop lines to the line A B, Fig. 60. Then, with C for a centre, carry these lines across the plan of the board, as seen at 1, 2, 3, 4, 5. Then take the

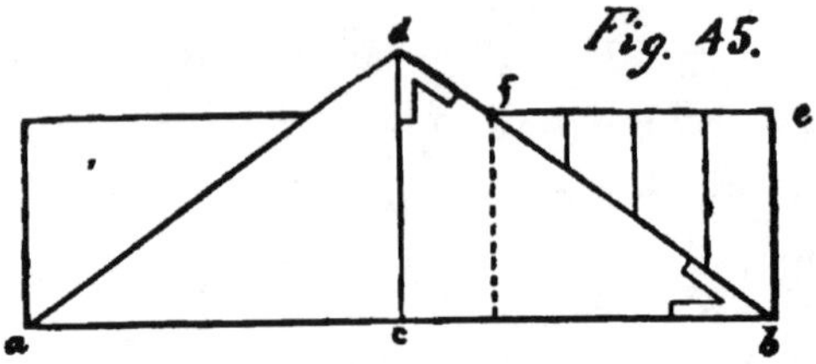

Fig. 45.

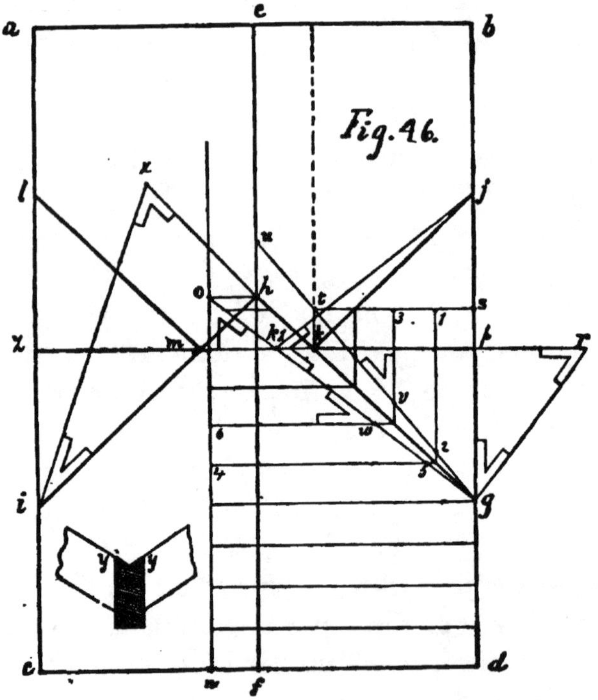

Fig. 46.

width of the board on the plan at 1, and lay it off at 1 *a*. Take the width on the plan at 2, and lay it off at 2 *a*, and so on. Then draw a curved line from *c* through these points to *a*, and also from *c* through these points to *b*. The result gives the shape of the boards.

Plate 25 shows the method of finding the rake moulding to fit any gutter; also the method of finding the level moulding to fit the rake. Fig. 61 represents the gutter, Fig. 62 is the rake moulding, and Fig. 63 is the level moulding. To find the shape of the rake moulding: From the gutter to be used, saw off a piece half an inch long; lay this piece on a smooth board, and mark around it, as seen in Fig. 61. Then, from the upper, outer point of the gutter, draw the line *a a*, giving the pitch of the roof. Then through several prominent points in the outline of the gutter, draw lines parallel to the first line, as *b b*, *c c*, etc. Then from the face of the fillet draw the vertical line A B. Then for the rake draw the line A B, Fig. 62, at right angles with the pitch of the roof. Take the distance from *a* to the line A B, Fig. 61, measured square across, as indicated by the dotted line, and lay it off from the line A B, Fig. 62, parallel with the slant of the roof. Do the same with the other points. Then draw a line through these points, which gives the shape of the rake moulding. Then, to get the shape of the level moulding, take the distance from *a* to the line A B, Fig. 61, and lay it off square from the line A B, Fig. 63, as indicated by the dotted lines. Do the same with the other points. Then draw a line through these points, which gives the shape of the level moulding.

Plate 26 shows mitre boxes for rake mouldings. Fig. 64 shows a box with cuts for mitring the rake moulding to the gutter. The angle across the top of the box is the mitre. (See Plate 36.) The angles on the sides of the box are the same as the down bevel at the top of the rafters. In sawing, keep nearest you the side of the boxes shown in the cut. Place the moulding upside down in the box, keeping the moulded side toward you, as shown in Fig. 65; taking care to have the bevel of the moulding at *c* fit well against the side of the box. Let *a b*, Fig. 64, represent a piece of rake moulding; cut the mitre at *a*, in the end of the box just above it, letting the moulding lay the same as the line *a b*. The mitre on the end *a* will fit the mitre of the gutter on the right-hand side of the gable. Cut the mitre at *d*, in the end of the box just above *d*, holding the moulding as before described. The mitre on the end *d* will fit the mitre of the gutter on the left side of the gable. To mitre the rake mouldings together at

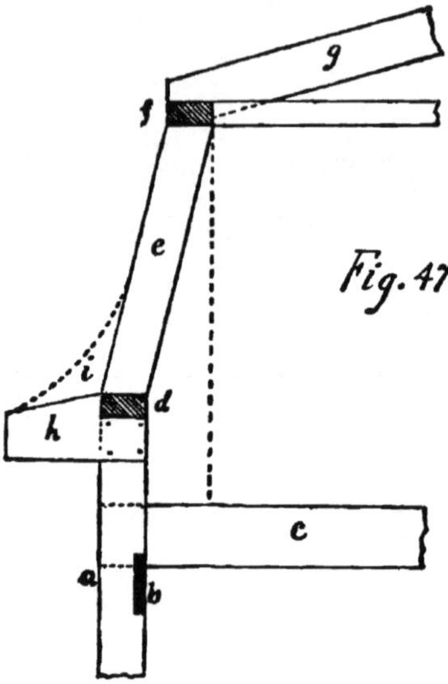

Fig. 47.

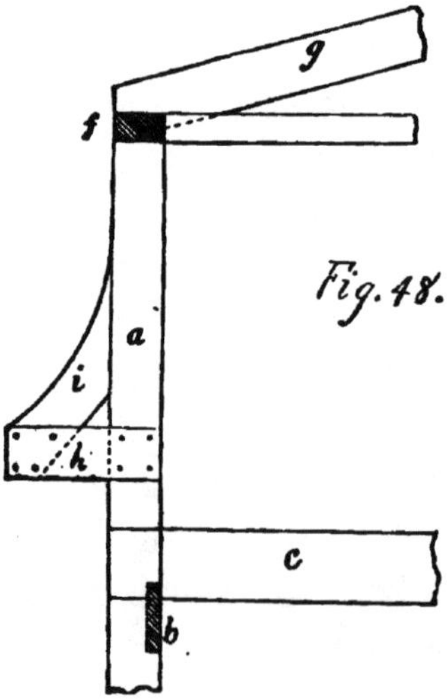

Fig. 48.

the top, the box shown in Fig. 66 is used. The angles on the top of the box are the same as the down bevel at the top of the rafters, the sides being sawed down square. Place the moulding in the box, as shown in Fig. 67, keeping the bevel at *c* flat on the bottom of the box, and having the moulded side toward you. The moulding *a b*, Fig. 64, is turned end for end, which brings it the other edge up, *a b*, Fig. 66; and the mitre for the top is cut on the end *b* in the end of the box just above it, which completes the moulding for the right-hand side of the gable. The mitre for the top of the moulding for the left side of the gable is cut on the end *c* of the moulding *c d*, in the end of the box just above *c*.

When the rake moulding is made of the proper form, these boxes are very convenient; but a great deal of the machine-made mouldings are not of the proper form to fit the gutter. In such cases, the moulding should be altered to the proper form if they come very bad; although many use the mouldings as they come, and trim the mitres so as to make them do.

Plate 27. Fig. 68 represents a plan of a flight of stairs, with a wind at the top. Plate 28, Fig. 69, is a more detailed plan of the wind; and Plate 29, Fig. 71, is an elevation of the winding posts showing the position of the mortises and risers. The lettering on these different cuts is the same for each part of the work: *b* is the face stringer, or carriage; *a* is the newel post; *c* is the winding post; *e* is the post at the upper landing, and is cut away so as to hook on to the upper floor, as seen in Fig. 69; *d* is a short piece of stringer, connecting the two posts *c* and *e; f* is the skirting-board, which is fastened to the trimmer, and makes a finish of the well-room. The risers 1 and 4 are tenoned into the post *c*, as shown in Fig. 68. Suppose, for instance, that the rise is seven inches: then the top of riser 1 is seven inches above the mortise, for the face-stringer *b*. (See also Fig. 71.) The top of riser 4 is twenty-one inches above riser 1. The top of riser 4, and the piece of stringer *d*, are even. The top of riser 5 is seven inches above *d*, or riser 4. The width of the winding steps are alike, when measured on a circle, struck from the winding post *c*, as shown in Fig. 68 at 1, 2, 3, 4. The face and centre stringers are usually made of two-inch plank. The wall-stringer is often made of a good stout inch board. The winding risers are made four or five inches wider than the others; the extra width projecting below the preceding riser, so as to afford a good nailing for the pieces of plank, 1 *a*, 2 *a*, 3 *a*, sometimes called chocks, and the piece of stringer *d* 2. The bottom step is frequently, as it is in this case, made a couple of inches wider than the rest of

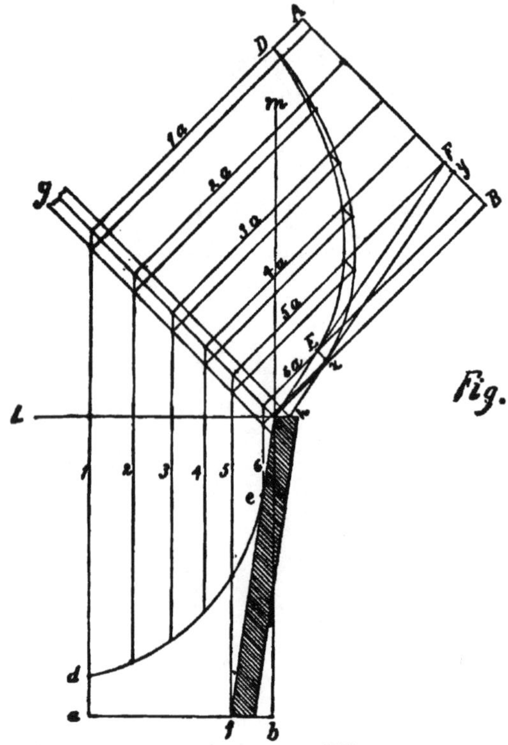

Fig. 49.

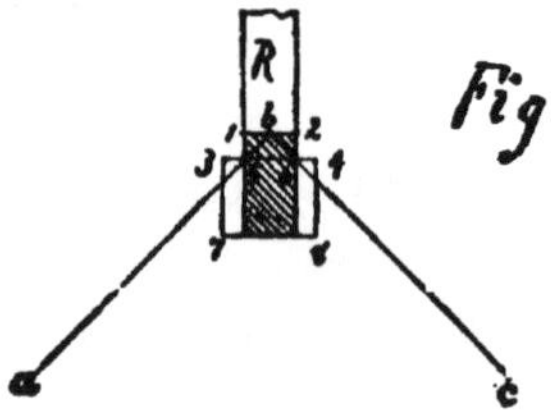

Fig. 50.

the steps. The risers are mitred into the face-stringer (and in laying out the face-stringer do not forget to allow for this), and are grooved to receive the tongue of the steps, as seen in Plate 30, Fig. 72. The ends of the steps are returned on the face-stringer, and a scotia moulding is mitred around beneath. The steps and risers are generally grooved to receive the base, which is tongued to fit; but a very cheap flight of stairs might be built with the wall-stringer nailed to the base, the steps and risers being butted against the base.

Fig. 70 shows the manner of laying out a stair-stringer, by taking the width of the step on the blade of the square, and taking the rise on the tongue of the square; *r* being the risers, and *s* being the steps. Steps will generally finish three-fourths of an inch wider than the width of step on the stringer.

Plate 30. Fig. 73 shows how to find the length of opening in the floor, to give sufficient head-room for the stairs. Suppose that the story is 9 feet in the clear, and the upper flooring, lathing, and plastering, etc., is 13 inches: then the stairs must be 9 feet + 13 inches = 10 feet 1 inch, from top to bottom, that is, 121 inches. Now, if we assume 7 inches for the rise, we have $17\frac{2}{7}$ risers. Since we must have a whole number of risers, we will adopt 17 as the number of risers, then the exact width of riser is $121 \div 17 = 7\frac{2}{17}$ inches, practically, $7\frac{1}{8}$ inches. We will make our steps 9 inches on the stringer; but they will finish nearly an inch wider, owing to their projecting beyond the riser.

Now the rise being $7\frac{1}{8}$ inches, we find that when we have ascended 3 risers, that is, $21\frac{3}{8}$ inches, we have 7 feet $2\frac{5}{8}$ inches head-room. Now counting out from the top of the stairs, we find that this point is the width of 14 steps from the top, which is 14×9 inches $=$ 126 inches $=$ 10 feet 6 inches. So with an opening of 10 feet 6 inches, we have 7 feet $2\frac{5}{8}$ inches head-room. If we can do with less head-room, we ascend another riser, which takes us up $28\frac{1}{2}$ inches, leaving us still a head-room of 6 feet $7\frac{1}{2}$ inches; this is at a point the width of 13 steps from the top, which is 13×9 inches $=$ 117 inches $=$ 9 feet 9 inches: so that, with an opening 9 feet 9 inches in length, we still have 6 feet $7\frac{1}{2}$ inches head-room. The opening might be still further reduced in length, if necessary, by narrowing the steps an inch or so.

We have said nothing about hand-rails, as there are firms of stair-builders in every large city who can furnish rails, posts, and balusters — by sending them a sketch of the stairs (similar to Plate 27, in this book), giving the width of the staircase, and the width of the riser and step, measured on the

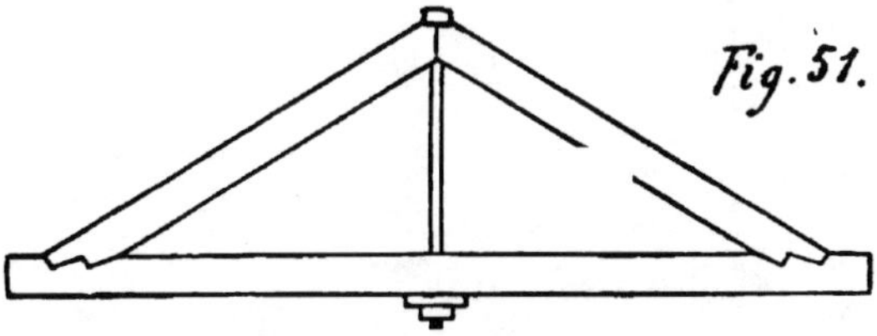

Fig. 51.

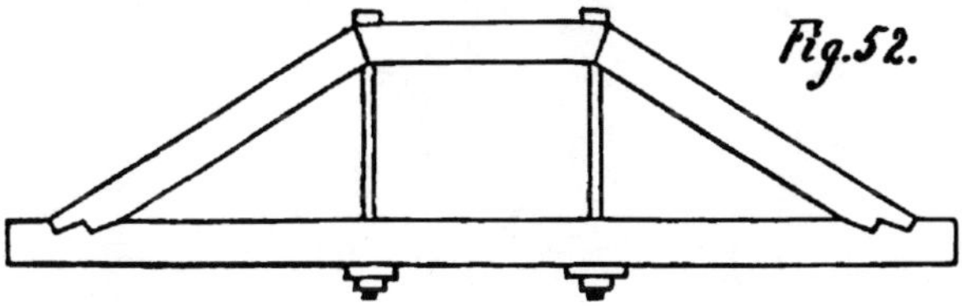

Fig. 52.

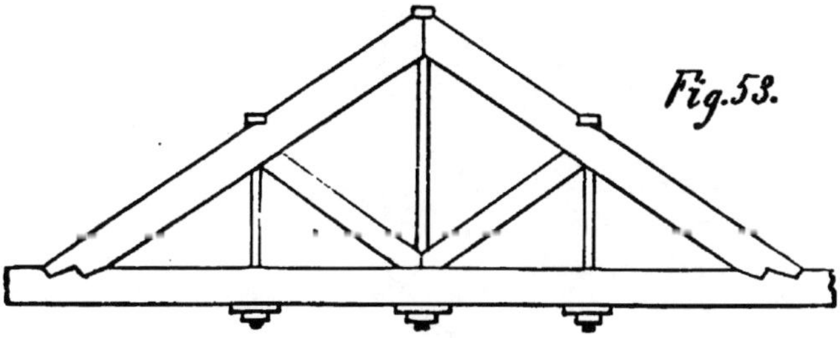

Fig. 53.

Plate 21.

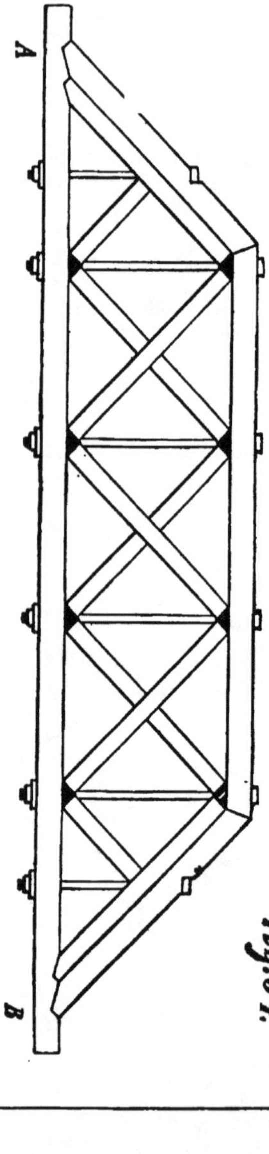

Fig.54.

stringer — a great deal cheaper than an inexperienced man can make them.

Plate 31. Fig. 74 shows a method of eight-squaring a stick of square timber. Lay a two-foot square or rule on the side of the timber, keeping both ends of one edge even with the edges of the timber; mark off at 7 and 17 inches, gauge off on all sides of the timber the distance in that these points come from the edge, and it gives the lines to hew by. At *a* is shown the end of the timber.

A board may be divided into any number of equal parts in a similar manner. If it is desired to divide the board into 10 equal parts, have the corner of the square even with one edge of the board, and have 20 inches come even with the other edge; then mark off every 2 inches. If 7 parts were desired, make 21 inches even with the other edge, instead of 20 inches, and mark off every 3 inches instead of every 2 inches.

Figs. 75 and 76 illustrate a very simple and also a very accurate method of fitting down thresholds. Take any board 5 or 6 inches wide, as *b*, Fig. 75, and 2 or 3 inches longer than the width of the doorway *d*. Lay this board on the floor, keeping the edge of the board one inch from the door-frame; lay a short straight-edge (2 feet square) against the door-jamb, and mark on the board where it crosses; also lay it against the rebate of the jamb, and mark on the board where it crosses; repeat the operation on the other door-jamb. Now draw back this board, and substitute the threshold in the place of the door-frame, keeping the upper corner of the threshold one inch from the edge of the board, as seen in the shaded section in Fig. 76; and continue the lines from the board *b* on to the threshold *t*. Now all that remains to be done is to gauge on to the threshold the depth of the rebate. If carefully done, the threshold will be a perfect fit every time. A hard pencil sharpened fine, or, better still, a knife, should be used in marking.

Fig. 77 represents a round chimney or flagstaff, *a*, passing through a slanting roof: the shape of the opening in the roof will be oval, as shown at *c*.

Plate 32 shows the mitring of straight and circular mouldings. Fig. 78 shows four circular mouldings, mitred together so as to form one moulding, as shown at A. The centres of all these mouldings come together at *a*. The mitre joint where 1 and 2 come together is a straight line, *b a*. The mitre where 2 and 4 come together is a curved line, one end of which is at the intersection of the edges of the moulding at *c;* the other end is at the intersection of the centre lines at *a;* the amount of curvature is found by the intersection of lines *e* and *f*, running midway be-

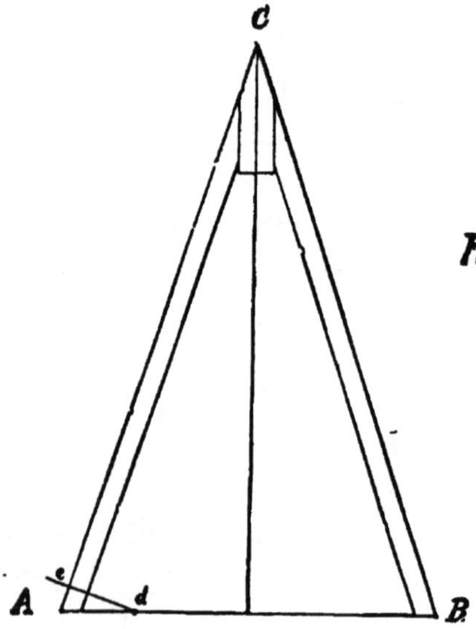

Fig. 55.

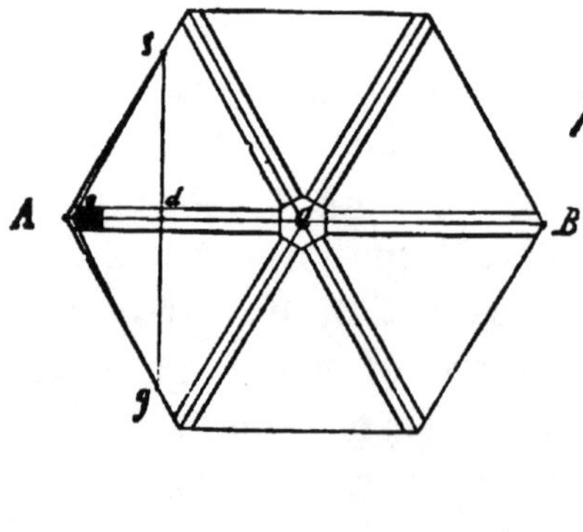

Fig. 56.

tween the centres and the outsides of these mouldings, the intersections being at *d*. Now, with these three points, *a*, *d*, and *c*, find the centre of a circle, the circumference of which will pass through them. (See Fig. 9, Plate 3.) Figs. 79 and 81 are other illustrations of the mitring of straight and circular mouldings. The intersection of the outsides at *a* and *c*, and the intersection of the centre lines at *b*, give three points, with which find a centre as before. Fig. 80 shows a wide and a narrow strip of board mitred together: the intersection of the outsides gives the angle of the mitre. It should be remarked that the mouldings in Figs. 78 and 79 must be the same shape each side of the centres, such as are called double mouldings.

Plate 33 shows the method of finding the bevels for a hopper-box having butt joints. Fig. 82: *a b c d* is the plan of the top of the box, and *e f g h* is the plan of the bottom. Fig. 83: A B is the line of the bottom of the box; *a b* is the slant of the sides, which line continue indefinitely toward *c*. Draw the line *d b* at right angles with *a b*. At any point on the line A B, as at *f*, drop a perpendicular line until it intersects the line *a c*. Now with *f* as a centre, and a radius tangent to the line *a c*, cut the line A B at *h;* join *h c;* at *h* is the bevel to cut the sides. Then, again, with *f* as a centre, and a radius tangent to *b d*, cut the line

A B at *g*. Join *g c;* at *g* is the bevel to cut the edges, the stock being jointed square on the edges.

Plate 34 shows the method of finding the bevels for a hopper-box with mitre joints. Fig. 84: *a b c d* is the plan of the top of the box, and *e f g h* is the plan of the bottom. Fig. 85: Let *a b c* be one corner of the plan of the box, and *b d* be the slant of the side. Draw the line *d f* at right angles with the slant of the side *b d*. Bisect the angle *a b c*, getting the line *b e*, which would be the mitre for the edges if the sides were perpendicular; but as the sides slant, the correct mitre is found by erecting a perpendicular on the line *b c*, as at *h*, continuing it until it intersects the line *b e*. Now, with *h* as a centre, and a radius tangent to *b f*, cut *b c* at *g*. Join *g e*. At *g* is the mitre for the edge. Then with *h* as a centre, and a radius tangent to *b d*, cut the line *b c* at *c*. Now join *c e*. At *e* we have the bevel to cut the sides.

Plate 35. Fig. 86 is an elevation of a splayed circular-top window. Fig. 87 shows the method of finding the form of a board to spring around the splayed circular top on the inside, the principle being the same as finding the envelope of a truncated cone (See Plate 7, Fig. 21); the bevel of the sides being continued till they intersect at *a*, which is the point to use as a centre, to describe the form of the board.

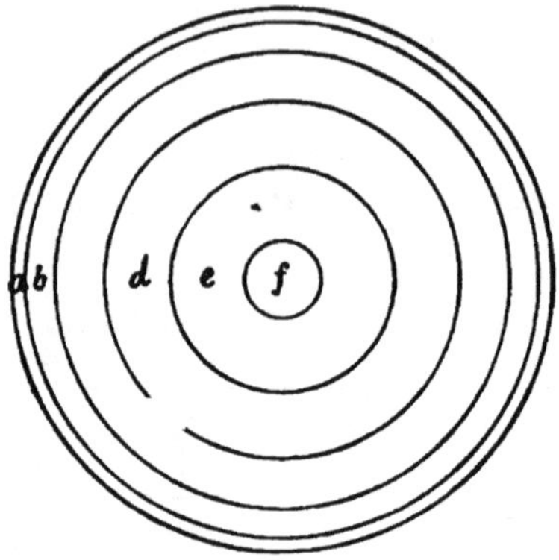

Fig. 57.

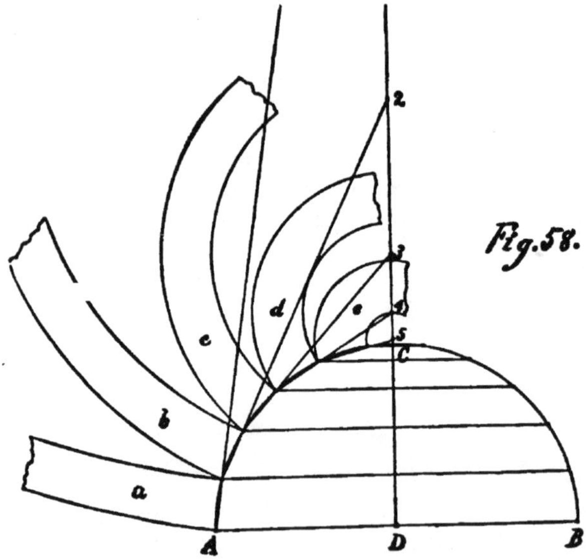

Fig. 58.

BUILDER'S ESTIMATES.

Stone-work is estimated by the perch; $24\frac{3}{4}$ cubic feet making one perch. An 18-inch wall, 1 foot high, and $16\frac{1}{2}$ feet long, contains one perch.

Brick-work. — Bricks are usually estimated at 25 to the cubic foot. They usually lay 5 courses to each foot in height.

For an 8-inch wall, allow 17 bricks for each square foot of surface. For a 12-inch wall, allow 25 bricks for each square foot of surface. For a 16-inch wall, allow 34 bricks for each square foot of surface.

Chimneys.

SIZE OF CHIMNEY.	NO. OF FLUES.	SIZE OF FLUES.	NO. OF BRICKS PER FOOT IN HEIGHT.
16 × 16 inches	1	8 × 8 inches	30
16 × 24 "	1	8 × 10 "	40
16 × 28 "	2	8 × 8 "	50
16 × 40 "	3	8 × 8 "	70
16 × 52 "	4	8 × 8 "	90
20 × 20 "	1	12 × 12 "	40
20 × 24 "	1	12 × 16 "	45

The above does not include waste, which must be allowed.

Mortar for Brick-work.—One cask of good lime to a load (about 20 bushels) of sand is sufficient for 1,000 or 1,100 bricks.

Cement for Cellar Bottoms should be mixed in the proportion of 1 of cement to 3 of gravel, and should be laid 3 inches thick. One cask of cement will cover 5 or 6 square yards.

Plasterers' Mortar.—One cask of lime to a load (20 bushels) of sand, and 2 bushels of hair, will cover about 50 square yards of surface; and ½ cask of lime will skim the same. In estimating the surface to be covered, plasterers deduct only half the area of openings, such as doors and windows, from the square yards in the walls.

TIMBERS.

Timbers for a Light Frame.—Sills, 4×6 or 6×6 inches. Flooring-timbers, 2×6 inches, put from 16 to 22 inches apart. Posts, 3×5 inches. Ledger-boards, 1×6 inches, well fitted and nailed. Studding, 2×3 inches, put 16 inches to centres. Plates, 3×4 inches. Rafters, 2×5 inches, put 2 feet apart. Partition studding, 2×3 and strapping 1×3 inches, put 16 inches to centres.

Timbers for a Medium Frame.—Sills, 6×7, 7×8, or 8×8 inches. Flooring timbers, 2×8, 9, or 10 inches,

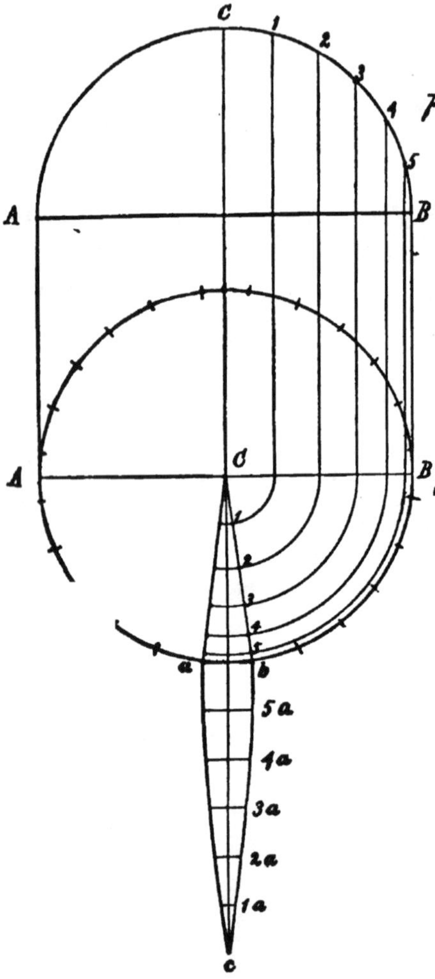

Fig. 59.

Fig. 60.

put 16 or 18 inches apart, and bridged. Posts, 4 × 6 or 4 × 8 inches. Studding, 2 × 4 inches, put 16 inches to centres. Window and door studs, 3 × 4, or 4 × 4 inches. Ledger-boards, 1 × 7 or 8 inches, well fitted and nailed, or girts 4 or 5 × 7 or 8 inches. Plates, 3 × 4 inches. Rafters, 2 × 6 inches, put 2 feet apart. Main partition studs 2 × 4 inches; other partitions, 2 × 3 inches, put 12 or 16 inches to centres.

Timbers for a Good Heavy Frame for Dwelling-House. — Sills, 8 × 8 or 8 × 10 inches. Flooring timbers, first story, 2 × 12 inches; second story, 2 × 10 inches; third story, 2 × 8 inches, put 16 or 18 inches apart, and well bridged. Side girts, 5 × 8 inches. End girts, 6 × 8 inches. Outside studding, 2 × 5 inches, put 12 or 16 inches to centres. Window and door studs, 3 × 5 inches. Rafters, 2 × 8 inches, put 20 or 24 inches apart. Main partitions, 2 × 5 inches; other partitions, 2 × 4 inches, put 12 or 16 inches to centres.

To square the sills of a house, make a mark on the upper outside edge of the side sill 8 feet from the corner of the house, and 6 feet from the corner of the house on the end sill; when the sills are square, a 10-foot pole will just reach across from point to point.

Framing and Boarding. — To estimate the number of square feet of boards required to board a building, and lay the under floors, double the length, and also the width of the building; add these amounts, which gives the length around the building; multiply this by the length of the outside studding, which gives the square feet in the walls of the house. If the house has a pitch roof, multiply the width of the house by the rise of the roof: the result will be the square feet in 2 gables. Then, to find the square feet in the roof: to the length of the house, add the amount of projection at both ends (generally about 18 inches at each end, which makes 3 feet to be added), which amount multiply by twice the total length of the rafters, which gives the square feet in the roof. Then for the floors, multiply the length of the building by the width, and multiply this by the number of floors, which gives the square feet in all of the floors. Add together these different amounts, and add $\frac{1}{8}$ for waste, which will give the number of square feet required. In estimating the labor in framing and boarding, some builders reckon eight or ten dollars per thousand feet.

TABLE OF BRACES.

RUN.	LENGTH OF BRACE.
2 ft. 3 in. × 2 ft. 3 in.	3 ft. 2$\frac{3}{16}$ in.
2 ft. 6 in. × 2 ft. 6 in.	3 ft. 6$\frac{7}{16}$ in.
2 ft. 9 in. × 2 ft. 9 in.	3 ft. 10$\frac{5}{8}$ $\frac{1}{32}$ in.
3 ft. 0 in. × 3 ft. 0 in.	4 ft. 2$\frac{7}{8}$ $\frac{1}{32}$ in
3 ft. 3 in. × 3 ft. 3 in.	4 ft. 7$\frac{1}{8}$ $\frac{1}{32}$ in
3 ft. 6 in. × 3 ft. 6 in.	4 ft. 11$\frac{3}{8}$ in.
3 ft. 9 in. × 3 ft. 9 in.	5 ft. 3$\frac{5}{8}$ in.
4 ft. 0 in. × 4 ft. 0 in.	5 ft. 7$\frac{7}{8}$ in.
4 ft. 3 in. × 4 ft. 3 in.	6 ft. 0$\frac{1}{8}$ in.
4 ft. 6 in. × 4 ft. 6 in.	6 ft. 4$\frac{3}{8}$ in.
4 ft. 9 in. × 4 ft. 9 in.	6 ft. 8$\frac{5}{8}$ in.
5 ft. 0 in. × 5 ft 0 in.	7 ft. 0$\frac{3}{16}$ $\frac{1}{32}$ in.
1 ft. 6 in. × 2 ft. 0 in.	2 ft. 6 in.
3 ft. 0 in. × 4 ft. 0 in.	5 ft. 0 in.

Plate 25.

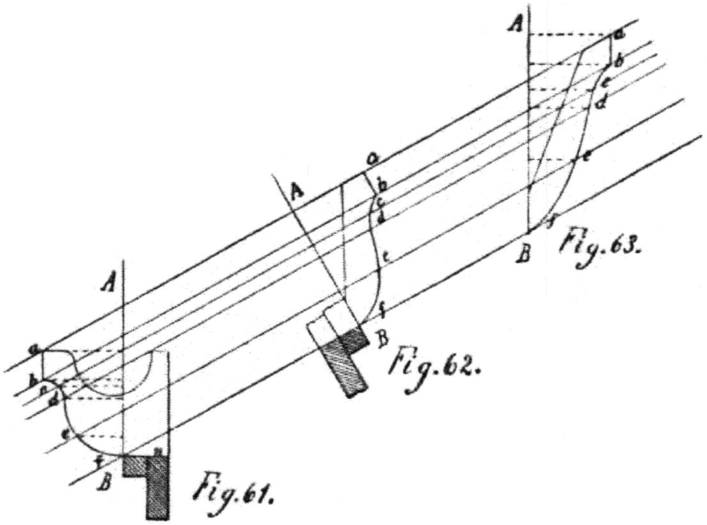

Fig. 61. Fig. 62. Fig. 63.

BOARD, PLANK, AND SCANTLING MEASURE.

Width.	1 In.	2 In.	3 In.	4 In. 2×2	5 In.	6 In. 2×3	7 In.	8 In. 2×4	9 In. 3×3
Length. Ft.	Ft. In.	Ft. In.	Ft. In.	Ft. In.	Ft. In.	Ft. In.	Ft. In.	Ft. In.	Ft. In.
1	0 1	0 2	0 3	0 4	0 5	0 6	0 7	0 8	0 9
2	0 2	0 4	0 6	0 8	0 10	1 0	1 2	1 4	1 6
3	0 3	0 6	0 9	1 0	1 3	1 6	1 9	2 0	2 3
3½	0 3	0 7	0 10	1 2	1 5	1 9	2 0	2 4	2 7
4	0 4	0 8	1 0	1 4	1 8	2 0	2 4	2 8	3 0
4½	0 4	0 9	1 1	1 6	1 10	2 3	2 7	3 0	3 4
5	0 5	0 10	1 3	1 8	2 1	2 6	2 11	3 4	3 9
5½	0 5	0 11	1 4	1 10	2 3	2 9	3 2	3 8	4 1
6	0 6	1 0	1 6	2 0	2 6	3 0	3 6	4 0	4 6
6½	0 6	1 1	1 7	2 2	2 8	3 3	3 9	4 4	4 10
7	0 7	1 2	1 9	2 4	2 11	3 6	4 1	4 8	5 3
7½	0 7	1 3	1 10	2 6	3 1	3 9	4 4	5 0	5 7
8	0 8	1 4	2 0	2 8	3 4	4 0	4 8	5 4	6 0
8½	0 8	1 5	2 1	2 10	3 6	4 3	4 11	5 8	6 4
9	0 9	1 6	2 3	3 0	3 9	4 6	5 3	6 0	6 9
9½	0 9	1 7	2 4	3 2	3 11	4 9	5 6	6 4	7 1
10	0 10	1 8	2 6	3 4	4 2	5 0	5 10	6 8	7 6
10½	0 10	1 9	2 7	3 6	4 4	5 3	6 1	7 0	7 10
11	0 11	1 10	2 9	3 8	4 7	5 6	6 5	7 4	8 3
11½	0 11	1 11	2 10	3 10	4 9	5 9	6 8	7 8	8 7
12	1 0	2 0	3 0	4 0	5 0	6 0	7 0	8 0	9 0
12½	1 0	2 1	3 1	4 2	5 2	6 3	7 3	8 4	9 4
13	1 1	2 2	3 3	4 4	5 5	6 6	7 7	8 8	9 9
13½	1 1	2 3	3 4	4 6	5 7	6 9	7 10	9 0	10 1
14	1 2	2 4	3 6	4 8	5 10	7 0	8 2	9 4	10 6
14½	1 2	2 5	3 7	4 10	6 0	7 3	8 5	9 8	10 10
15	1 3	2 6	3 9	5 0	6 3	7 6	8 9	10 0	11 3
15½	1 3	2 7	3 10	5 2	6 5	7 9	9 0	10 4	11 7
16	1 4	2 8	4 0	5 4	6 8	8 0	9 4	10 8	12 0
16½	1 4	2 9	4 1	5 6	6 10	8 3	9 7	11 0	12 4
17	1 5	2 10	4 3	5 8	7 1	8 6	9 11	11 4	12 9
17½	1 5	2 11	4 4	5 10	7 3	8 9	10 2	11 8	13 1
18	1 6	3 0	4 6	6 0	7 6	9 0	10 6	12 0	13 6
18½	1 6	3 1	4 7	6 2	7 8	9 3	10 9	12 4	13 10
19	1 7	3 2	4 9	6 4	7 11	9 6	11 1	12 8	14 3
19½	1 7	3 3	4 10	6 6	8 1	9 9	11 4	13 0	14 7
20	1 8	3 4	5 0	6 8	8 4	10 0	11 8	13 4	15 0

If it be desired to find the square feet in a board which is longer than 20 feet, take the square feet in two shorter boards, the added lengths of which are equal to the length of the board which you wish to measure; for instance, if the board be 26 feet long by 19 inches wide, add together the square feet in a 20-foot and a 6-foot board, each 19 inches wide: 31 feet 8 inches + 9 feet 6 inches = 41 feet 2 inches.

(OVER)

BOARD, PLANK, AND SCANTLING MEASURE—Continued.

Width.	10 In. 2×5		11 In.		12 In. 2×6 3×4		13 In.		14 In. 2×7		15 In. 3×5		16 In. 2×8 4×4		17 In.		18 In. 2×9 3×6	
Length. Ft.	Ft.	In.	Ft.	In.	Ft.	In.	Ft.	In.	Ft.	In.	Ft.	In.	Ft.	In.	Ft.	In.	Ft.	In.
1	0	10	0	11	1	0	1	1	1	2	1	3	1	4	1	5	1	6
2	1	8	1	10	2	0	2	2	2	4	2	6	2	8	2	10	3	0
3	2	6	2	9	3	0	3	3	3	6	3	9	4	0	4	3	4	6
3½	2	11	3	2	3	6	3	9	4	1	4	4	4	8	4	11	5	3
4	3	4	3	8	4	0	4	4	4	8	5	0	5	4	5	8	6	0
4½	3	9	4	1	4	6	4	10	5	3	5	7	6	0	6	4	6	9
5	4	2	4	7	5	0	5	5	5	10	6	3	6	8	7	1	7	6
5½	4	7	5	0	5	6	5	11	6	5	6	10	7	4	7	9	8	3
6	5	0	5	6	6	0	6	6	7	0	7	6	8	0	8	6	9	0
6½	5	5	5	11	6	6	7	0	7	7	8	1	8	8	9	2	9	9
7	5	10	6	5	7	0	7	7	8	2	8	9	9	4	9	11	10	6
7½	6	3	6	10	7	6	8	1	8	9	9	4	10	0	10	7	11	3
8	6	8	7	4	8	0	8	8	9	4	10	0	10	8	11	4	12	0
8½	7	1	7	9	8	6	9	2	9	11	10	7	11	4	12	0	12	9
9	7	6	8	3	9	0	9	9	10	6	11	3	12	0	12	9	13	6
9½	7	11	8	8	9	6	10	3	11	1	11	10	12	8	13	5	14	3
10	8	4	9	2	10	0	10	10	11	8	12	6	13	4	14	2	15	0
10½	8	9	9	7	10	6	11	4	12	3	13	1	14	0	14	10	15	9
11	9	2	10	1	11	0	11	11	12	10	13	9	14	8	15	7	16	6
11½	9	7	10	6	11	6	12	5	13	5	14	4	15	4	16	3	17	3
12	10	0	11	0	12	0	13	0	14	0	15	0	16	0	17	0	18	0
12½	10	5	11	5	12	6	13	6	14	7	15	7	16	8	17	8	18	9
13	10	10	11	11	13	0	14	1	15	2	16	3	17	4	18	5	19	6
13½	11	3	12	4	13	6	14	7	15	9	16	10	18	0	19	1	20	3
14	11	8	12	10	14	0	15	2	16	4	17	6	10	0	19	10	21	0
14½	12	1	13	3	14	6	15	8	16	11	18	1	19	4	20	6	21	9
15	12	6	13	9	15	0	16	3	17	6	18	9	20	0	21	3	22	6
15½	12	11	14	2	15	6	16	9	18	1	19	4	20	8	21	11	23	3
16	13	4	14	8	16	0	17	4	18	8	20	0	21	4	22	8	24	0
16½	13	9	15	1	16	6	17	10	19	3	20	7	22	0	23	4	24	9
17	14	2	15	7	17	0	18	5	19	10	21	3	22	8	24	1	25	6
17½	14	7	16	0	17	6	18	11	20	5	21	10	23	4	24	9	26	3
18	15	0	16	6	18	0	19	6	21	0	22	6	24	0	25	6	27	0
18½	15	5	16	11	18	6	20	0	21	7	23	1	24	8	26	2	27	9
19	15	10	17	5	19	0	20	7	22	2	23	9	25	4	26	11	28	6
19½	16	3	17	10	19	6	21	1	22	9	24	4	26	0	27	7	29	3
20	16	8	18	4	20	0	21	8	23	4	25	0	26	8	28	4	30	0

To reckon the square feet in a board, multiply the width in inches by the length in feet, and divide this result by 12, which gives the number of square feet it contains.

BOARD, PLANK, AND SCANTLING MEASURE—

Concluded.

Width.	19 In.	20 In. 2 × 10 4 × 5	21 In. 3 × 7	22 In. 2 × 11	23 In.	24 In. 2 × 12 3 × 8 4 × 6	25 In. 5 × 5	26 In. 2 × 13
Length. Ft.	Ft. In.	Ft. In.	Ft. In.	Ft. In.	Ft. In.	Ft. In.	Ft. In.	Ft. In.
1	1 7	1 8	1 9	1 10	1 11	2 0	2 1	2 2
2	3 2	3 4	3 6	3 8	3 10	4 0	4 2	4 4
3	4 9	5 0	5 3	5 6	5 9	6 0	6 3	6 6
3½	5 6	5 10	6 1	6 5	6 8	7 0	7 3	7 7
4	6 4	6 8	7 0	7 4	7 8	8 0	8 4	8 8
4½	7 1	7 6	7 10	8 3	8 7	9 0	9 4	9 9
5	7 11	8 4	8 9	9 2	9 7	10 0	10 5	10 10
5½	8 8	9 2	9 7	10 1	10 6	11 0	11 5	11 11
6	9 6	10 0	10 6	11 0	11 6	12 0	12 6	13 0
6½	10 3	10 10	11 4	11 11	12 5	13 0	13 6	14 1
7	11 1	11 8	12 3	12 10	13 5	14 0	14 7	15 2
7½	11 10	12 6	13 1	13 9	14 4	15 0	15 7	16 3
8	12 8	13 4	14 0	14 8	15 4	16 0	16 8	17 4
8½	13 5	14 2	14 10	15 7	16 3	17 0	17 8	18 5
9	14 3	15 0	15 9	16 6	17 3	18 0	18 9	19 6
9½	15 0	15 10	16 7	17 5	18 2	19 0	19 9	20 7
10	15 10	16 8	17 6	18 4	19 2	20 0	20 10	21 8
10½	16 7	17 6	18 4	19 3	20 1	21 0	21 10	22 9
11	17 5	18 4	19 3	20 2	21 1	22 0	22 11	23 10
11½	18 2	19 2	20 1	21 1	22 0	23 0	23 11	24 11
12	19 0	20 0	21 0	22 0	23 0	24 0	25 0	26 0
12½	19 9	20 10	21 10	22 11	23 11	25 0	26 0	27 1
13	20 7	21 8	22 9	23 10	24 11	26 0	27 1	28 2
13½	21 4	22 6	23 7	24 9	25 10	27 0	28 1	29 3
14	22 2	23 4	24 6	25 8	26 10	28 0	29 2	30 4
14½	22 11	24 2	25 4	26 7	27 9	29 0	30 2	31 5
15	23 9	25 0	26 3	27 6	28 9	30 0	31 3	32 6
15½	24 6	25 10	27 1	28 5	29 8	31 0	32 3	33 7
16	25 4	26 8	28 0	29 4	30 8	32 0	33 4	34 8
16½	26 1	27 6	28 10	30 3	31 7	33 0	34 4	35 9
17	26 11	28 4	29 9	31 2	32 7	34 0	35 5	36 10
17½	27 8	29 2	30 7	32 1	33 6	35 0	36 5	37 11
18	28 6	30 0	31 6	33 0	34 6	36 0	37 6	39 0
18½	29 3	30 10	32 4	33 11	35 5	37 0	38 6	40 1
19	30 1	31 8	33 3	34 10	36 5	38 0	39 7	41 2
19½	30 10	32 6	34 1	35 9	37 4	39 0	40 7	42 3
20	31 8	33 4	35 0	36 8	38 4	40 0	41 8	43 4

To reckon the square feet in timber, multiply the width in inches by the thickness, and this result by the length in feet. This result divided by 12 gives the number of square feet contained in the piece of timber.

Shingles. — A bundle of shingles, if full size, should have 25 courses on each end, and be 20 inches wide; or else have 22 courses on one end, and 23 courses on the other, and be 22 inches wide. Four such bundles contain 1,000 shingles, each supposed to be 4 inches wide. They are usually 16 inches long; sometimes in the nicest class of shingles, they come 18 inches long. It is poor economy to use an inferior quality of shingles; it costs rather more to lay them than it does good ones, and they make a leaky roof, almost from the first. Spruce shingles are used considerably by some, but are not suitable to make a good roof, as they warp and twist, and very quickly split to pieces. Some soft pine or cedar shingles, best quality, are the cheapest in the end: but even bundles of the best quality will contain some hard, glassy shingles, which will act almost as badly as spruce; they should be thrown out.

It takes about 5 pounds of four-penny nails per thousand shingles; or 3 or 4 pounds of three-penny coarse, which we think are preferable.

One thousand shingles, laid 4 inches to the weather, will cover 111 square feet. One thousand shingles, laid 4½ inches to the weather, will cover 125 square feet. One thousand shingles, laid 5 inches to the weather, will cover 139 square feet. One thousand shingles (18-inch shingles only, except on walls), laid 5½ inches to the weather, will cover 153 square feet.

The above does not include waste, which must be allowed.

Laths are 4 feet long, and come in bundles of 100 each.

Plate 26.

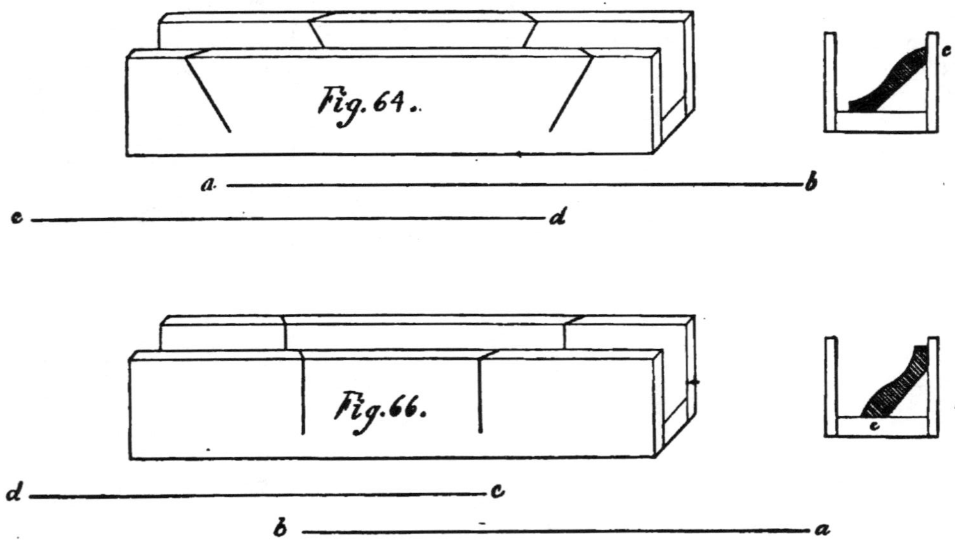

Fig. 64.

Fig. 65.

Fig. 66.

Fig. 67.

(We have seen some lots, the bundles of which were short some 20 or 30 laths.) Ten bundles make 1,000, which will cover about 60 square yards, which requires about 7 pounds of three-penny fine nails.

Clapboards are usually 4 feet long, and come 25 in a bundle; 4 bundles making a hundred, which requires about $3\frac{1}{2}$ pounds of five-penny nails. One hundred clapboards, laid 4 inches to the weather, will cover 133 square feet. One hundred clapboards, laid $4\frac{1}{2}$ inches to the weather, will cover 150 square feet. This does not include waste, which must be allowed.

Sandpaper. — No. 00, too fine. No. 0, too fine. No. $\frac{1}{2}$, fine enough for rubbing down paint or shellac. No. 1, fine for carpenters. No. $1\frac{1}{2}$, generally used. No. 2, too coarse.

Sheet Lead and Zinc for Flashings. — Sheet lead $\frac{1}{32}$ inch thick weighs 2 pounds per square foot; $\frac{3}{64}$ inch thick, weighs 3 pounds per square foot (generally used); $\frac{1}{16}$ inch thick, weighs 4 pounds per square foot; $\frac{3}{32}$ inch thick, weighs 6 pounds per square foot; $\frac{1}{8}$ inch thick, weighs 8 pounds per square foot. Sheet zinc comes in sheets 3×7 feet. A sheet of No. 9 zinc (commonly used) weighs 14 pounds, that is, about $\frac{2}{3}$ pounds per square foot.

To bend a Gooseneck. — Fill the lead pipe full of sand, ram it in well and plug up both ends, bend it carefully over your knee, or around a barrel or smooth tree.

WINDOWS.

These are the sizes made by Boston door, sash, and blind manufacturers.

12 Lights.	4 Lights.	Width.	Height.	Weight, 1¼ in. Thick.	Weight, 1⅜ in. Thick.	Weight, 1½ in. Thick.	Line for each Weight.
Inches.	Inches.	Ft. In.	Ft. In.	Lbs.	Lbs.	Lbs.	Feet.
7× 9	10½×18	2 0⅝	3 5	3	–	–	2¼
–	10½×22	2 0⅝	4 1	–	–	–	2¾
8×10	12 ×20	2 3⅝	3 9	3¼	3½	4¼	2½
8×12	12 ×24	2 3⅝	4 5	4	4¼	4½	3
–	12 ×25	2 3⅝	4 7	–	–	–	3¼
9×11	13½×22	2 6⅝	4 1	–	–	4½	2¾
9×12	13½×24	2 6⅝	4 5	4¼	4½	4¾	3
9×13	13½×26	2 6⅝	4 9	4½	4¾	5	3¼
9×14	13½×28	2 6⅝	5 1	4¾	5	5¼	3½
9×15	13½×30	2 6⅝	5 5	–	–	5½	3¾
9×16	13½×32	2 6⅝	5 9	–	–	6	4
–	14 ×26	2 7⅝	4 9	–	4¾	5	3¼
–	14 ×28	2 7⅝	5 1	–	5	5¼	3½
–	14 ×30	2 7⅝	5 5	–	5¼	5½	3¾
10×12	15 ×24	2 9⅝	4 5	–	–	5	3
10×13	15 ×26	2 9⅝	4 9	–	–	5½	3¼
10×14	15 ×28	2 9⅝	5 1	–	5½	5¾	3½
10×15	15 ×30	2 9⅝	5 5	–	5¾	6	3¾
10×16	15 ×32	2 9⅝	5 9	–	6¼	6½	4
10×17	15 ×34	2 9⅝	6 1	–	–	6½	4¼
10×18	15 ×36	2 9⅝	6 5	–	–	6¾	4½
–	10 ×28	2 11⅝	5 1	–	–	6¼	3½
–	16 ×30	2 11⅝	5 5	–	–	6½	3¾
–	16 ×32	2 11⅝	5 9	–	–	6¾	4
–	16 ×34	2 11⅝	6 1	–	–	7	4¼
12×15	18 ×30	3 3⅝	5 5	–	–	7	3¾
12×16	18 ×32	3 3⅝	5 9	–	–	7½	4
–	18 ×34	3 3⅝	6 1	–	–	8	4¼
12×18	18 ×36	3 3⅝	6 5	–	–	8½	4½
12×20	18 ×40	3 3⅝	7 1	–	–	9	5

Pockets in window-frames should be cut from 15 to 20 inches in length, according to the size of the frame. For the slant of the window sill, see explanations accompanying Fig. 22, Plate 8.

WINDOWS — Concluded.

2 Lights.	Width.	Length.	Weig't, 1½ in thick.	Line each weight.	2 Lights.	Width.	Length.	Weig't, 1½ in. thick	Line each weight.
Inches.	Ft. In.	Ft. In.	Lbs.	Ft.	Inches.	Ft. In.	Ft. In.	Lbs.	Ft.
13½ × 24	1 5	4 5	3	3	16 × 32	1 7½	5 9	4½	4
13½ × 26	1 5	4 9	3¼	3¼	16 × 34	1 7½	6 1	4¾	4¼
13½ × 28	1 5	5 1	3½	3½	18 × 26	1 9½	4 9	4	3¼
13½ × 30	1 5	5 5	4	3¾	18 × 28	1 9½	5 1	4¼	3½
13½ × 32	1 5	5 9	4½	4	18 × 30	1 9½	5 5	4½	3¾
15 × 24	1 6½	4 5	3¼	3	18 × 32	1 9½	5 9	4¾	4
15 × 26	1 6½	4 9	3½	3¼	18 × 34	1 9½	6 1	5	4¼
15 × 28	1 6½	5 1	3¾	3½	18 × 36	1 9½	6 5	5¼	4½
15 × 30	1 6½	5 5	4	3¾	20 × 28	1 11½	5 1	4	3½
15 × 32	1 6½	5 9	4¼	4	20 × 30	1 11½	5 5	4¼	3¾
15 × 34	1 6½	6 1	4½	4¼	20 × 32	1 11½	5 9	5	4
15 × 36	1 6½	6 5	4¾	4½	20 × 34	1 11½	6 1	5¼	4¼
16 × 28	1 7½	5 1	4	3½	20 × 36	1 11½	6 5	5½	4½
16 × 30	1 7½	5 5	4¼	3¾					

CELLAR WINDOW SASH.

Size of Glass.	Thickness.	No. of Lights.	Height.	Size of Glass.	Thickness.	No. of Lights.	Height.
6 × 8 in.	1¼ in.	3 lights.	1 lt. high	9 × 13 in.	1¼ in.	4 lights.	2 lt. high
6 × 8 "	1¼ "	4 "	1 "	9 × 14 "	1¼ "	3 "	1 "
6 × 8 "	1¼ "	4 "	2 "	9 × 15 "	1¼ "	3 "	1 "
7 × 9 "	1¼ "	2 "	1 "	9 × 16 "	1½ "	3 "	1 "
7 × 9 "	1¼ "	3 "	1 "	9 × 17 "	1½ "	3 "	1 "
7 × 9 "	1¼ "	4 "	1 "	9 × 18 "	1½ "	3 "	1 "
7 × 9 "	1¼ "	4 "	2 "	10 × 8 "	1¼ "	3 "	1 "
8 × 10 "	1¼ "	2 "	1 "	10 × 12 "	1¼ "	3 "	1 "
8 × 10 "	1¼ "	3 "	1 "	10 × 13 "	1¼ "	3 "	1 "
8 × 10 "	1¼ "	4 "	1 "	10 × 14 "	1¼ "	3 "	1 "
8 × 10 "	1¼ "	4 "	2 "	10 × 15 "	1¼ "	3 "	1 "
8 × 12 "	1¼ "	3 "	1 "	10 × 16 "	1¼ "	3 "	1 "
8 × 12 "	1¼ "	4 "	1 "	10 × 17 "	1½ "	3 "	1 "
8 × 12 "	1¼ "	4 "	2 "	10 × 18 "	1½ "	3 "	1 "
9 × 11 "	1¼ "	3 "	1 "	11 × 15 "	1½ "	3 "	1 "
9 × 12 "	1¼ "	3 "	1 "	11 × 16 "	1½ "	3 "	1 "
9 × 12 "	1¼ "	4 "	1 "	11 × 17 "	1½ "	3 "	1 "
9 × 12 "	1¼ "	4 "	2 "	11 × 18 "	1½ "	3 "	1 "
9 × 13 "	1¼ "	3 "	1 "	12 × 16 "	1½ "	3 "	1 "
9 × 13 "	1¼ "	4 "	1 "	12 × 18 "	1½ "	3 "	1 "

Blinds are the same width as windows, and are one-half inch longer.
There are usually about 75 feet of line in a hank.
Skylights or Hotbed Sash. — Outside measures, 3 feet × 6 feet; 3 feet × 5 feet; 2½ feet × 4 feet; 2¼ feet × 3½ feet; 2 feet × 3 feet.

NOTE. — Frames for cellar-window sash. For 1 light high, and for 2 lights wide, make frame 3½ inches larger than glass. For 2 lights high, and for 3 lights wide, make frame 3¾ inches larger than glass.

Bins for Grain. — A Winchester bushel is 18½ inches in diameter by 8 inches deep, and contains 2,150.42 cubic inches, nearly 2,150½ cubic inches, and is used for measuring fine stuff like grain, beans, etc.

To estimate the size of a box or bin to hold a certain number of bushels, multiply the number of cubic inches in one bushel by the number of bushels which the bin is to hold: this will give the number of cubic inches which the bin will contain. Now assume any two of the three dimensions of the bin, say the length and the width; multiply the number of inches in length by the number of inches in width; divide the number of cubic inches to be contained in the bin by this product: the result will be the number of inches in depth of the bin. A cubic foot contains about ⅘ of a bushel.

A box 9 inches × 9 inches × 6½ inches deep will contain 1 peck.

A box 12 inches × 12 inches × 7½ inches deep will contain ½ bushel.

A box 14 inches × 14 inches × 11 inches deep will contain 1 bushel.

5-bushel box or bin: 30 inches × 30 inches × 12 inches deep, or 25 inches × 25 inches × 17$\frac{3}{16}$ inches deep.

10-bushel bin: 30 inches × 30 inches × 24 inches deep, or 2 feet × 3½ feet × 21$\frac{5}{16}$ inches deep, or 3½ feet × 3½ feet × 12$\frac{3}{16}$ inches deep.

15-bushel bin: 3½ feet × 3½ feet × 18¼ inches deep, or 3 feet × 4 feet × 18⅝ inches deep.

20-bushel bin: 3½ feet × 3½ feet × 24⅜ inches deep, or 3½ feet × 4 feet × 21$\frac{5}{16}$ inches deep, or 3 feet × 4 feet × 24⅞ inches deep.

25-bushel bin: 3 feet × 4 feet × 31⅛ inches deep, or 3½ feet × 4½ feet × 23$\frac{11}{16}$ inches deep, or 3 feet × 5 feet × 24⅞ inches deep.

30-bushel bin: 3½ feet × 4½ feet × 28½ inches deep, or 3 feet × 5 feet × 29⅞ inches deep.

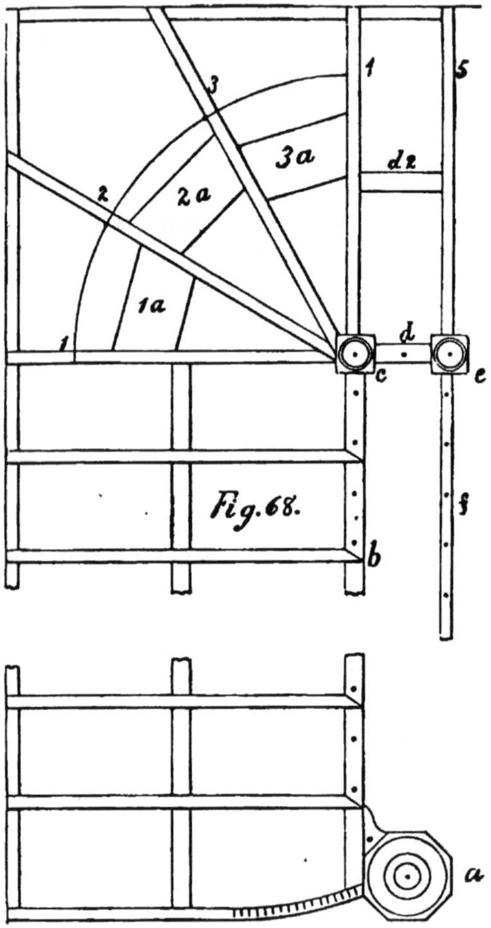

Fig. 68.

40-bushel bin: 4 feet × 5 feet × $29\frac{3}{4}$ inches deep, or 4 feet × 6 feet × $24\frac{7}{8}$ inches deep.

50-bushel bin: 4 feet × 6 feet × $31\frac{1}{8}$ inches deep, or $4\frac{1}{2}$ feet × 7 feet × $23\frac{11}{16}$ inches deep, or 5 feet × 6 feet × $24\frac{7}{8}$ inches deep.

A common flour-barrel will hold about $3\frac{1}{3}$ bushels of grain or other fine stuff.

Bins for Apples, Potatoes, etc. — In measuring coarse stuff, like apples, potatoes, etc., the bushel is heaped so that the cone, formed by the stuff being heaped, shall be not less than 6 inches in height. A heaped bushel contains 2,747.7 cubic inches, about $2,747\frac{3}{4}$ cubic inches.

5-bushel box or bin: 30 inches × 30 inches × $15\frac{1}{4}$ inches deep, or 2 feet × 3 feet × $15\frac{3}{4}$ inches deep.

10-bushel bin: $2\frac{1}{2}$ feet × $3\frac{1}{2}$ feet × $21\frac{3}{4}$ inches deep, or 3 feet × 4 feet × 16 inches deep.

15-bushel bin: 3 feet × 4 feet × $23\frac{7}{8}$ inches deep.

20-bushel bin: 3 feet × 4 feet × 32 inches deep, or $3\frac{1}{2}$ feet × 4 feet × $27\frac{1}{4}$ inches deep.

25-bushel bin: $3\frac{1}{2}$ feet × 4 feet × 34 inches deep, or 3 feet × 5 feet × $31\frac{3}{4}$ inches deep, or $3\frac{1}{2}$ feet × 5 feet × $27\frac{1}{4}$ inches deep.

30-bushel bin: 3 feet × 5 feet × 38 inches deep, or $3\frac{1}{2}$ feet × 5 feet × $32\frac{3}{4}$ inches deep.

40-bushel bin: $3\frac{1}{2}$ feet × 6 feet × $36\frac{1}{4}$ inches deep, or 4 feet × 6 feet × $31\frac{3}{4}$ inches deep.

50-bushel bin: 4 feet × 6 feet × $39\frac{3}{4}$ inches deep, or 5 feet × 5 feet × $38\frac{1}{8}$ inches deep, or 5 feet by 6 feet × $31\frac{3}{4}$ inches deep.

A common flour-barrel will hold about $2\frac{1}{2}$ bushels of apples or potatoes.

To estimate the Size of a Tank to hold a certain Number of Gallons. — A gallon contains 231 cubic inches. A cubic foot contains about $7\frac{1}{2}$ gallons. Multiply the number of cubic inches in one gallon by the number of gallons, which will give the number of cubic inches which the tank will contain; now assume any two of the three dimensions of the tank, say the length and the breadth; multiply the number of inches in length by the number of inches in breadth; divide the number of cubic inches contained in the tank by this product: the result will be the number of inches in depth of the tank. A barrel contains $31\frac{1}{2}$ gallons.

To estimate the Size of a Bin to hold a certain Number of Tons of Coal. — A cubic foot of anthracite coal weighs from 50 to 55 pounds: so a ton will occupy a space of 36 or 40 cubic feet (36 cubic feet is usually considered correct). Multiply the number of tons which the bin is to contain by the number of cubic feet contained in one ton, which will give the number of cubic feet which the bin is to contain; assume any two of the three dimensions of the bin, say the length and breadth; multiply the length, in feet, by the breadth, also in feet; divide the number of cubic feet contained in the bin by this product: the result will be the depth of the bin in feet.

If a ton of coal occupies 36 cubic feet, then a bin 4 feet $\times$ $4\frac{1}{2}$ feet will hold a ton of coal for each 2 feet in depth; a bin 4 feet $\times$ 6 feet will hold a ton of coal for each 18 inches in depth; a bin 6 feet $\times$ 6 feet will hold a ton of coal for each 12 inches in depth.

Miscellaneous. — In painting, all knots and sappy places should have one or two coats of shellac varnish previous to the first coat of paint. In nice houses, the *entire woodwork* of the *inside* is given one or two coats of shellac previous to painting. This prevents the knots and sap from staining the paint yellow.

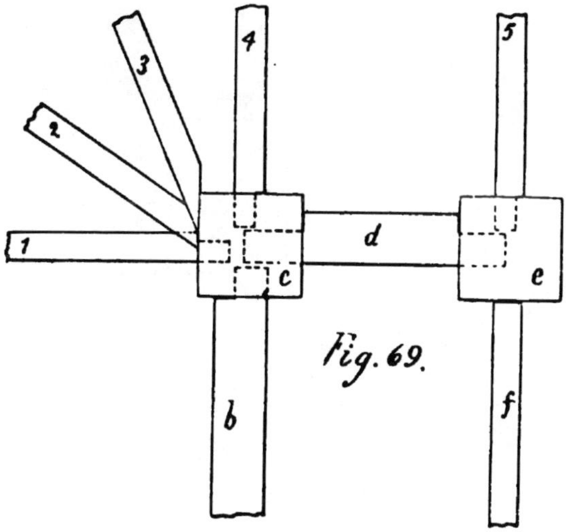

Fig. 69.

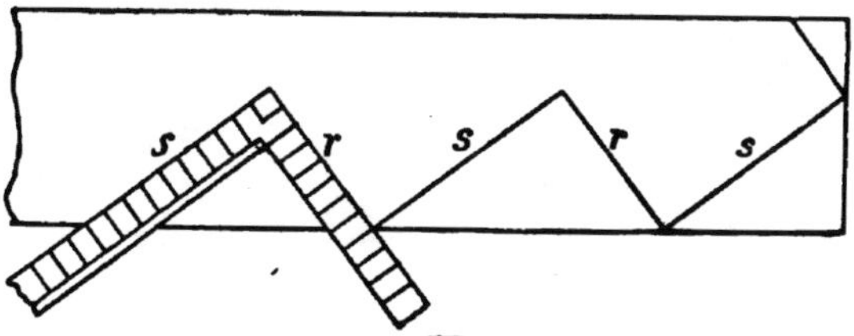

Fig. 70

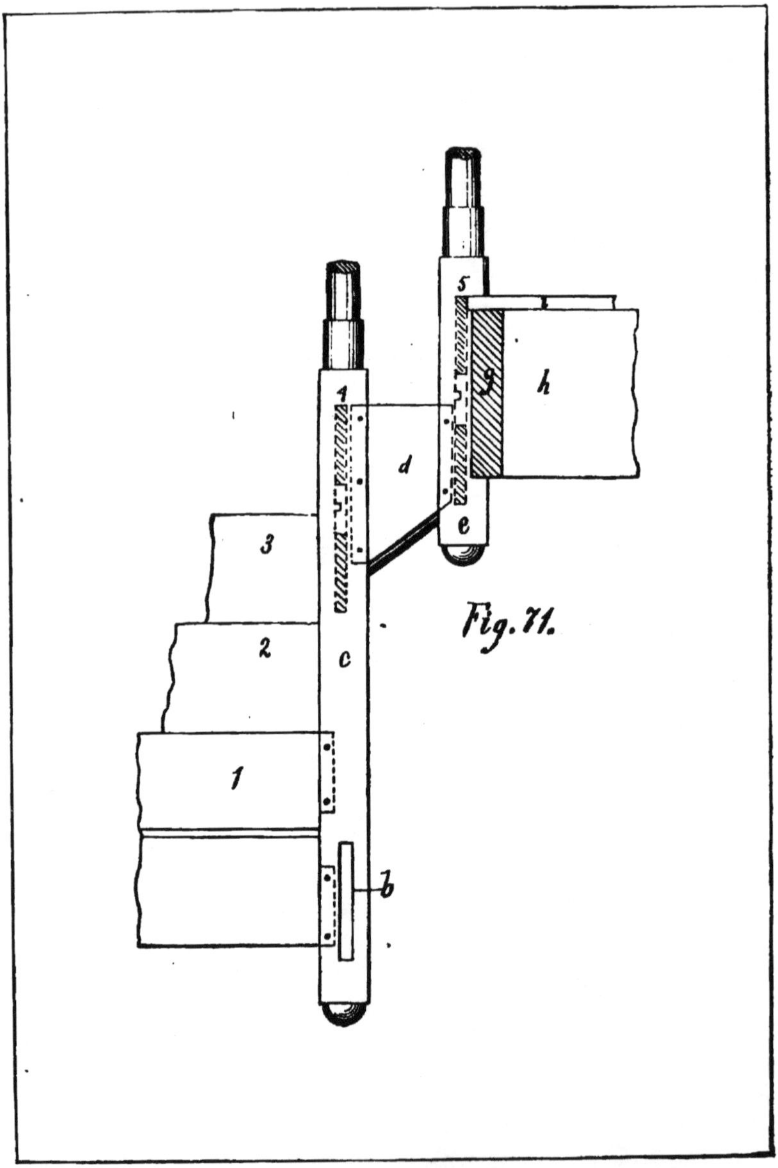

Fig. 71.

Window runs should never be varnished.

To help drawers and window sashes slide easily, rub the running parts with a piece of bayberry tallow or paraffine wax. This, however, is not a substitute for easing them with a plane.

A good thing to use in patching small scars in plastering is calcined plaster (sometimes called plaster-of-Paris), mixed with common flour paste. If the plaster is mixed with water it sets almost instantly; but when mixed with paste it sets quite slowly, giving time to use it as may be desired.

A flour-barrel is 28 to 30 inches high, and 20 or 21 inches in diameter at the largest part. This note may be of use in fitting up closets and pantries.

Weights of Various Materials.—These are taken from various sources, and are generally considered as practically correct, although different pieces of the same material will vary considerably: especially is this true of wood; one piece of dry pine will sometimes weigh nearly double as much as another. The weights given are per cubic foot, except when otherwise stated:—

Ash, 43 to 50 lbs; Babbitt metal, 456.32 (cubic inch, .263); beech, 43; birch, 37 to 44; brick and mortar, 115; boxwood, 80; cast brass, 537.75 (cubic inch, .31); cedar, 35; chalk, 145 to 162; charcoal, 18; chestnut, 38; cork, 15; cast copper, 537.3 (cubic inch, .31); cannel coal, 79.5; bituminous coal, 45 to 55; anthracite coal, 50 to 55; grindstone, 133.93; granite, 180; ebony, 74; English elm, 34 to 36; freestone, 150; flint glass, 192 (cubic inch, .111); crown and common green glass, 158 (cubic inch, .091); plate glass, 172 (cubic inch, .099); hornbeam, 47; cast iron, 451 (cubic inch, .26); wrought iron, 485 (cubic inch, .281); iron-wood, 71; ivory, 114; lignumvitæ, 83; cast lead, 708.5 (cubic inch, .41); sheet lead, 711.6; marble, 145 to 170; mercury, 848 (cubic inch, .49); Honduras mahogany, 35; Nassau mahogany, 42; Spanish mahogany, 53; maple, 42; white oak, 45 to 50; live oak, 70; white pine, 27 to 34; yellow pine, 32 to 40; rubber, 58; spruce, 29; silver, 653.8 (cubic inch, .377); steel, 499 (cubic inch, .288); dry sand, 117; sandstone, 140; water, 62.5; sea water, 64.18; cast zinc, 437 to 450 (cubic inch, .25); gold, 1,203 lbs. 10 ounces.

To distinguish Right-hand from Left-hand Loose Butts.—Take one in your hands, and open it so that the side having the countersunk holes for the screws will be up; then draw it apart, having the pintle pointing from you; then, if the part containing the pintle is in your right hand, it is a right-hand butt; if it is in your left hand, it is a left-hand butt. (See Plate 36. Fig. 91 shows a right-hand loose butt drawn apart, and Fig. 92 shows a left-hand loose butt.) The part of the butt containing the pintle belongs on the door-jamb, or door-frame. Right-hand butts go on right-hand doors, left-hand butts on left-hand doors. A door opening from you to the right is a right-hand door: one opening from you to the left is a left-hand door.

Plate 30.

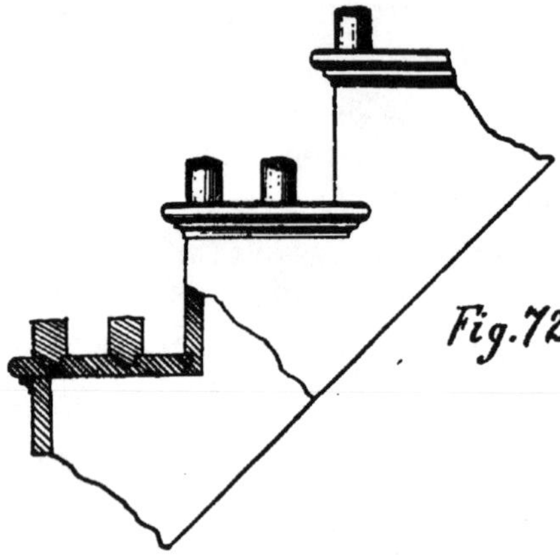

Fig. 72.

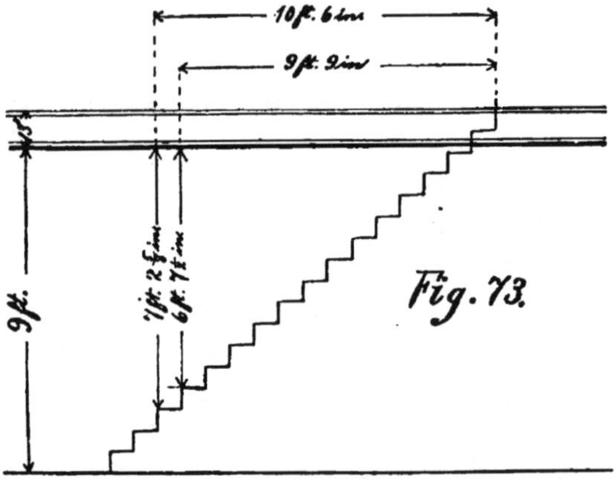

Fig. 73.

To find the Proper Angle to cut the Mitre of a Rake-moulding Mitre-box. — If the building is square, or has square corners, the mitre for the rake-moulding will be an angle of 45° let fall perpendicularly, when the moulding sets at the same slant as the roof. If the building is not square, then the angle for each corner of the building may be found by bisecting the angle formed by the side and end of the building (see Plate 1, Fig. 2) : then the mitre for the rake-moulding will be the angle, found as above, let fall perpendicularly. Set a bevel to the angle found, and mark the angle on the top of the mitre-box, as shown in Plate 36, Fig. 90, *a b*, representing the angle : then draw a line square from *b* to *c*. (If the building is square, the distance from *a* to *c* will be equal to the width of the box from outside to outside, from *b* to *c*.) In Fig. 89 we have shown the mitre-box set at the same slant as the roof, so that the angle on the side of the box stands perpendicularly; then lay off *a c* at right angles with *a b*, making the length of *a c* the same as the length of *a c* in Fig. 90; then draw the line *c d* with the bevel used to draw the down bevel at *a b* (which is the same as the down bevel of the rafters) ; then, to lay out the mitre on the box, make *a c*, Fig. 90, the same length as *a d*, Fig. 89 ; then square across from *c* to *b*, join *a* and *b*, which gives the actual angle to cut the mitre, so that, if the building is square, an angle of 45° let fall perpendicularly would describe this angle on the box, when the box is set on the same slant as the roof, as shown in Fig. 89. For convenience of workmen, we have laid out the mitres to cut rake-moulding mitre-boxes. They are as follows, and bevels can be set to the required number of degrees by the use of the protractor on Plate 8 : —

The angle of the mitre for $\frac{1}{3}$ pitch is about 40°.
The angle of the mitre for $\frac{2}{3}$ pitch is about 37°.
The angle of the mitre for square pitch is about 35°.

These are the angles for square buildings, and will not answer for other than square buildings.

Given the Diameter, to find the Rise for any Chord or Span.—It sometimes occurs, that it is desired to describe a segment of a circle of great radius; but the amount of rise is not known. For example: A building of brick or stone is to be constructed on part of a street which is curving; the radius of the curve being, say, 150 feet. The stone-cutter wants a pattern made to use for shaping the underpinning, the window-sills, etc.: he wants the pattern 8 feet long, so as to do for all the stone-work. Now, here we have an 8-foot segment of a 300-foot circle. It is impossible to make any thing like a true curve of that size by means of a line used as a radius. If we knew the amount of rise, we could describe the curve by means of the triangular frame described in Plate 3, Fig. 8; but, although the amount of rise is not known, it is a very easy matter to figure it out. The rule is as follows: Subtract the square of the chord or span from the square of the diameter, and extract the square root of the remainder. Subtract this root from the diameter, and halve the remainder, which gives the rise. To illustrate. The diameter being 300 feet, the square of the diameter is 300 × 300 feet, which is 90,000 feet. The square of the chord or span (8 feet) is 8 × 8 feet, which is 64 feet, which, subtracted from the square of the diameter, leaves 89,936 remainder, the square root of which is

Plate 31.

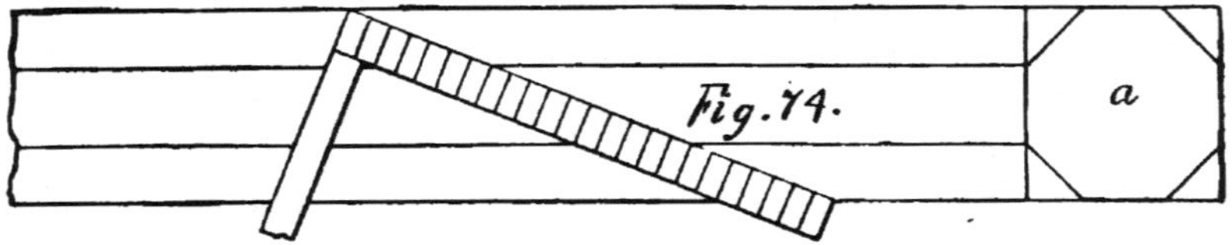

Fig. 74.

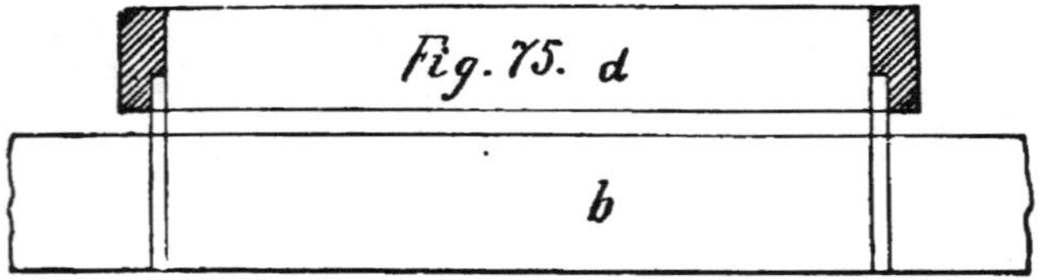

Fig. 75. d

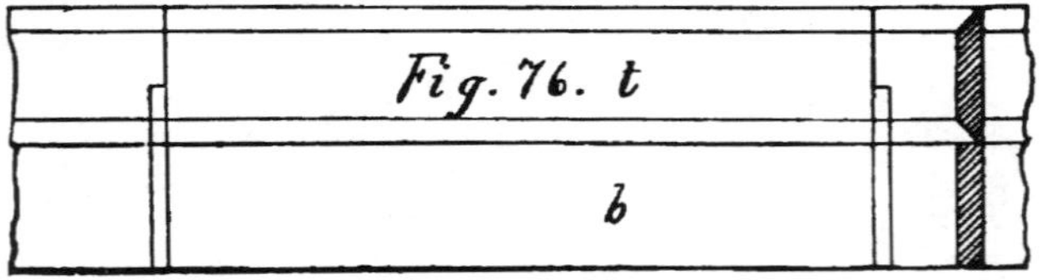

Fig. 76. t

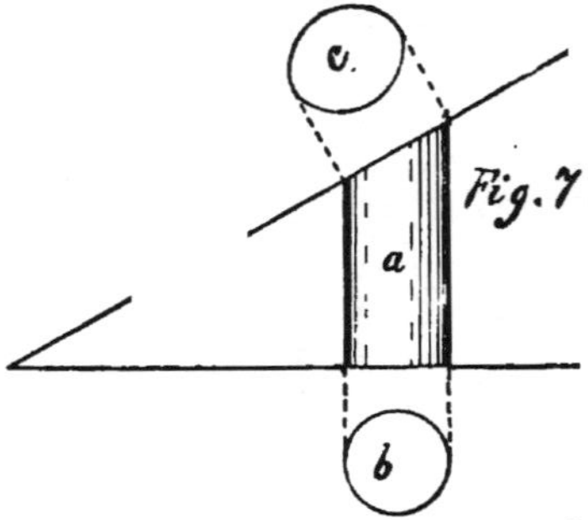

Fig. 77

299.893 +, which, subtracted from the diameter (300 feet), leaves .107 remainder, half of which gives .0535 feet as the rise, which we multiply by 12 to get the number of inches, which gives .642 inches. By referring to our table of decimal parts of an inch with fractional equivalents, we find that this is practically $\frac{5}{8}$ of an inch rise. Now, knowing the rise, and the chord or span, we can describe the curve by means of the frame arrangement described in Plate 3, Fig. 8.

Descriptions and Uses of the Various Markings on Rules and Squares, including the Slide-rule, and how to use it.— Although the markings on rules and squares were made for the express convenience of workmen, yet but very few understand the uses of them. Every workman ought to be perfectly familiar with all of them, so as to avail himself of every advantage they afford. On Plate 37 will be found illustrations of the most important markings. Fig. 93 is the board-measure commonly found on the back of the blade of ordinary 2-foot squares. To find the number of square feet in a board, find the number representing the length of the board in feet in the column under 12 inches, then in the same line find the number of square feet under the number of inches in width. For instance: Suppose a board is 14 feet long, and 6 inches wide, in the column under 12 inches we find 14, the length of the board in feet: then on the same line, under 6 inches (the width), we find 7, which is the number of square feet contained in the board. Again, suppose the board is 8 feet long, and $5\frac{1}{2}$ wide, under 12 inches we find 8 (the length of the board in feet): then on the same line we find that

$5\frac{1}{2}$ comes $\frac{2}{3}$ of the space beyond 3, which shows that the board contains $3\frac{2}{3}$ square feet, or 3 feet 8 inches. If the board is 12 feet long, then the number of inches in width will be the number of square feet contained in the board; or, if the board is 12 inches wide, then the number of feet in length will be the number of square feet contained in the board. Instead of finding the length in feet in the column under 12 inches, we may find the inches in width, then in the same line we will find the number of square feet under the number of inches that the board is feet in length. For instance: Suppose the board is 16 feet long, and 9 inches wide, under 12 inches we find 9, the width in inches: then under 16 inches, which represents 16 feet (the length), we find 12, which is the number of square feet which the board contains. If either the length or the width exceeds the figures on the square, find the square feet in a board of half the length or half the width, and double the result. Some say that this kind of board-measure is not exact, that it only *approximates*. This statement is not true. The whole number of square feet is found *exactly* where it occurs. For instance: A board 8 inches wide, which will contain 5 square feet, must be exactly $7\frac{1}{2}$ feet long; and it will be seen, by an inspection of the square, that the 5 occurs exactly under $7\frac{1}{2}$ inches on the square. It is not *approximate:* it is EXACT.

Fig. 94 exhibits what is called "The Essex Board-measure," which is adopted by some makers. In this style of board-measure the number of square feet and inches, or square feet and twelfths, are found under every inch in length of the square. The number of square feet

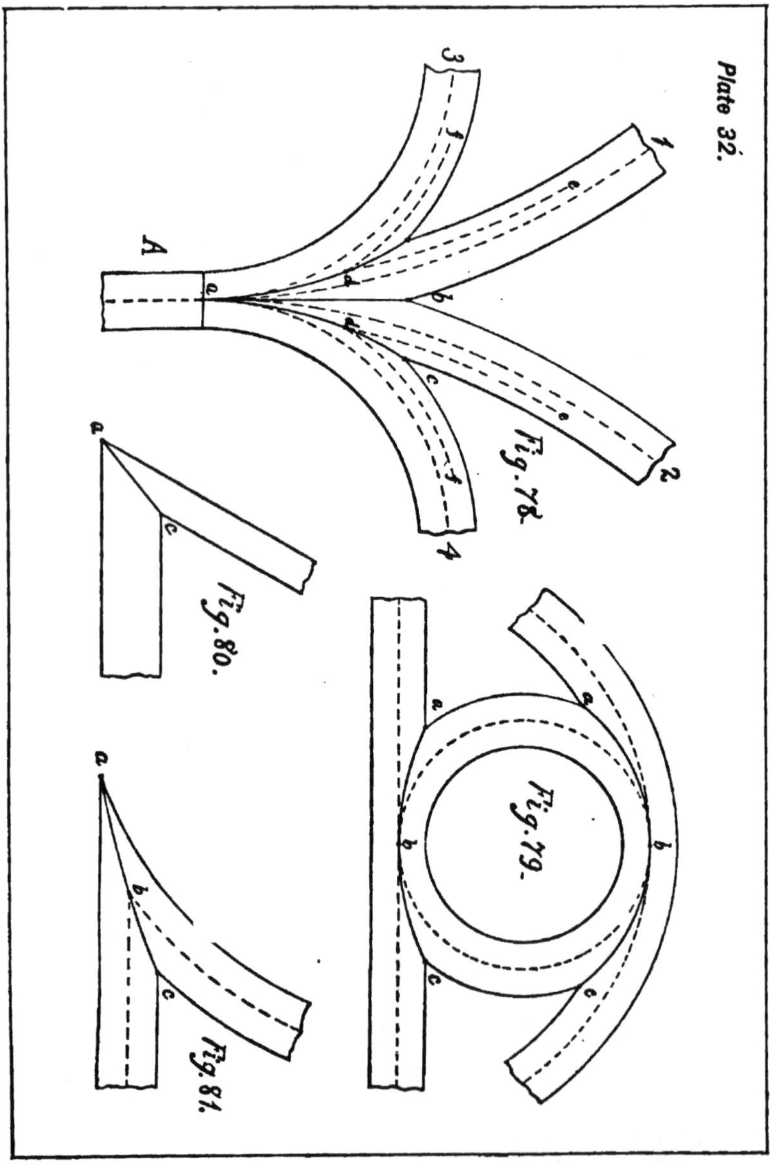
Plate 32.
A
a
b
c
d
e
f
1
2
3
4
Fig. 78.
Fig. 79.
Fig. 80.
Fig. 81.

in a board is found in the same manner with this kind of board-measure as with the former kind. Suppose a board is 10 inches wide, and 14 feet long, under 12 inches we find 14 (the length in feet) : then in the same line, under 10 inches (the width), we find 11–8, which represents $11\frac{8}{12}$ square feet.

Fig. 95 shows the brace-measure, which is marked on the tongue of squares. The two numbers at the left, one above the other, represent the runs in inches. The number and decimal at the right is the length of the brace in inches and hundredths. Thus, where the run is 57 by 57 inches, the length of the brace is 80.61 inches.

Fig. 96 shows the octagonal scale, which is found on the tongue of 2-foot squares. This scale is used to work from *centre lines.* If we desire to 8-square a stick of 10-inch timber, we first centre the width of each side, then, with a pair of compasses, take the distance from division 1 to division 10, and set it off on each side of the centre line; and by laying out both ends, and snapping a line, we have a guide to hew by. If the timber is 15 inches square, we take the distance from division 1 to division 15, and set it off on each side of the centre line as before. In many cases we are obliged to work from centre lines; as, for instance, when we 8-square a log, preparatory to rounding it, as in the case of mast and spar making, after having four sides flat, there is no corner to gauge from.

Fig. 97 shows the octagonal scales usually found on rules. The scale marked M is the same as the octagonal scale found on 2-foot squares, only it is sub-divided finer, and works from the centre in the same manner. The scale marked E works from the edge or corner. If a stick of timber is 12 inches square, we gauge on from the edge the distance from division 1 to division 12 on the scale E; or, if it is 14 inches square, we gauge on from the corner the distance from division 1 to division 14.

Fig. 98 is a draughting-scale, full size, with six different scales marked off on it. The first one is $\frac{1}{4}$ inch to the foot, or $\frac{1}{4}$ inch = 1 foot: then comes $\frac{1}{2}$ inch to the foot, $\frac{3}{4}$ inch to the foot, 1 inch to the foot, and also $1\frac{1}{4}$ and $1\frac{1}{2}$ inches to the foot. The first foot of the scale $\frac{1}{4}$ inch to the foot is divided into 6 parts, each part representing 2 inches. All the other scales have the first foot divided into inches. In using these scales to draw by, we begin to count the number of feet from the second foot, which is numbered 1, and count to the right: then, to get inches, we count to the left. For instance: If we are drawing with the scale of $\frac{3}{4}$ inch to the foot, and we want to get 2 feet 5 inches, we set one point of a pair of compasses to 5 inches, counting from the right hand toward the left, then extend the other point of the compasses to the right till it reaches the 2 feet, which gives us the required 2 feet 5 inches, which we transfer to our drawing. These scales are usually scattered around when put on to the 3-jointed rules, but on single-jointed rules they are often all put together the same as seen in Fig. 98.

Fig. 99 represents a scale of degrees, which is found on several draughting implements. To use this scale, we first draw a horizontal line (*a b*) a couple of inches long (see Plate 36, Fig. 88); set one point of a pair of compasses at the left-hand end (*a*) of this line, and, using the distance from C to 60 for a radius, describe part of a

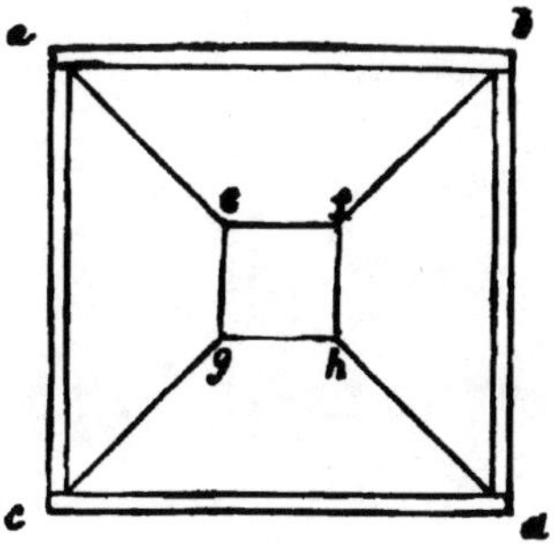

Fig. 82.

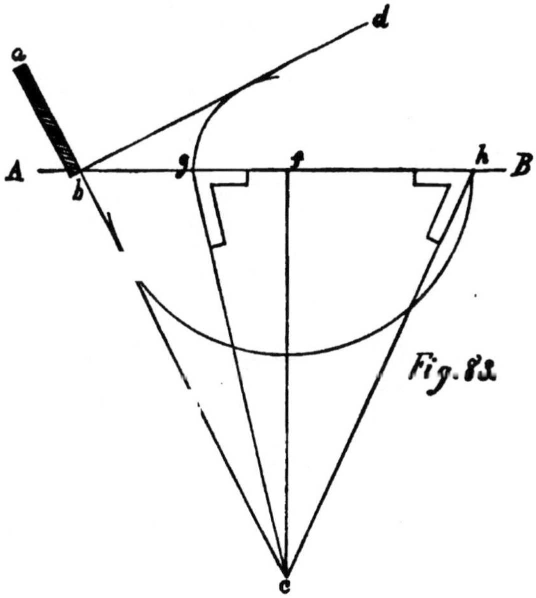

Fig. 83.

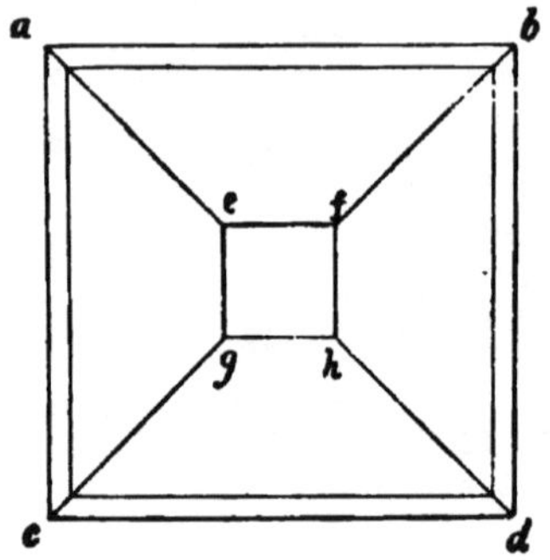

Fig. 84.

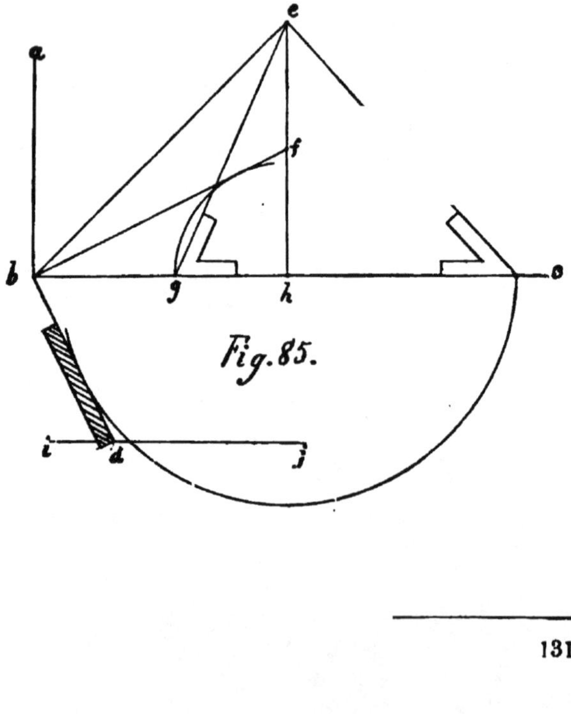

Fig. 85.

circle ($b\,c$); then using b as a centre, and with a radius equal to the length from C to any number of degrees desired, cut the segment $b\,c$, as seen at d; draw a line joining a and d, and we have an angle of the desired number of degrees.

Fig. 100 shows a diagonal scale for obtaining hundredths of an inch, which is found on some 2-foot squares, and on some draughting-scales. It is merely 1 square inch, divided vertically into 10 parts by horizontal lines. The upper and lower edges are divided into ten parts each: then a line is drawn from the upper left-hand corner to the first division on the bottom edge, another line from the first division at the upper edge to the second division on the lower edge, and so on. The space between the vertical line at the left, and the first diagonal line, is $\frac{1}{100}$ of an inch on the first line down from the top; each space to the right on this line is $\frac{10}{100}$ more; so that from the vertical line to the second diagonal line is $\frac{11}{100}$ on this same line, to the third diagonal line is $\frac{21}{100}$, and so on. From the vertical line to the first diagonal line is $\frac{2}{100}$ of an inch on the second line down from the top, and every space to the right on this line is $\frac{10}{100}$ more; so that from the vertical line to the second diagonal line is $\frac{12}{100}$, and so on. From the vertical line to the first diagonal line is $\frac{3}{100}$ on the third line down from the top, $\frac{4}{100}$ on the fourth line down, $\frac{5}{100}$ on the fifth line down, and so on. To mark off any number of inches and hundredths, measure off the desired number of inches, less 1 and the decimal, then, with a pair of compasses, take 1 inch and the required number of hundredths, and add it to the length already measured off. For instance: If we want to measure off

35.58 inches, we first measure off 34 inches; then on the eighth line down from the top, from the vertical line to the first diagonal line, is $\frac{8}{100}$, then to each of the others is $\frac{10}{100}$ more; so we take five of these spaces, which, with the first space, makes $\frac{58}{100}$; so we set a pair of compasses from the preceding inch to this point, and add it to the 34 inches already marked off, and it gives us the desired 35.58 inches.

THE SLIDE-RULE. FIG. 101.

The slide-rule consists of four lines, viz., A, B, C, D; A being on the upper edge of the rule, B being on the upper edge of the slide, C being on the lower edge of the slide, and D being on the lower edge of the rule. The lines A and B work together, and the lines C and D work together. The divisions and numbers on A and B are exactly alike; and, when closed, they stand thus: —

A	1	2	3	4	5	etc.
B	1	2	3	4	5	etc.

But, if 1 on the slide B is set to 2 on the rule A, then the numbers will stand thus: —

A	1	2	4	6	8	etc.
B		1	2	3	4	etc.

It will be seen that the proportion of 2 to 1 runs throughout, each number on A being the product of the number immediately underneath, on B, multiplied by 2; or, inversely, each number on B being the result of dividing the number immediately above, on A, by 2.

If 1 on the slide B is set to 3 on the rule A, the numbers will stand thus: —

A	1	3	6	9	12	etc.
B		1	2	3	4	etc.

Fig. 86.

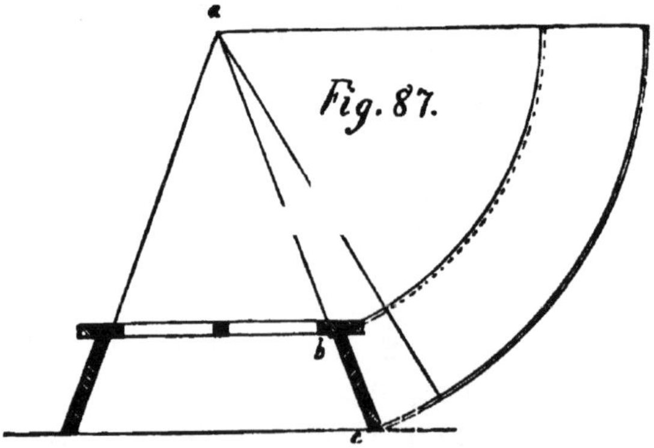

Fig. 87.

It will be seen that the proportion $\frac{3}{1}$ runs throughout, each number on A being the product of the number immediately underneath, on B, multiplied by 3; or, inversely, each number on B being the result of dividing the number immediately above, on A, by 3.

The C and D lines are relatively different, each number on the slide C being the square or self multiple of the number immediately underneath, on the rule D; or, inversely, each number on D being the square root of each number immediately above it, on C.*

The numbers and divisions are to be read decimally; for the spaces are, or are supposed to be, divided and subdivided into tens and tenths. The ordinary reading of the divisions on the lines A, B, and C, is, beginning at the left, 1, 2, 3, 4, 5, 6, 7, 8, 9, 10, which is marked 1; 11, which is not numbered; 12; then the intermediate numbers, 13, 14, 15, etc., which are not numbered, up to 20, which is marked 2; then the intermediate numbers, 21, 22, 23, etc., up to 30, which is marked 3; then continuing on to 40, which is marked 4; 50, which is marked 5; 60, marked 6; 70, marked 7; 80, marked 8; 90, marked 9; and 100, which is marked 10. Between 1 and 2 are 10 principal divisions, which indicate 1 plus any number of tenths. The first principal division beyond 1 indicates $1\frac{1}{10}$, the second division indicates $1\frac{2}{10}$, and so on up to 2. Each of these principal divisions between 1 and 2 are subdivided into 5 parts, each part representing $\frac{2}{100}$: so

* The square of any number is the result obtained by multiplying that number by itself: thus the square of 2 is $2\times2=4$; the square of 3 is $3\times3=9$. The square root of any number is that number which, when multiplied by itself, will produce the given number: thus the square root of 4 is 2, since $2\times2=4$; the square root of 9 is 3, since $3\times3=9$.

the first division beyond 1 is $1\frac{2}{100}$, the second is $1\frac{4}{100}$, the fifth is $1\frac{10}{100}$, or $1\frac{1}{10}$; the division next to 2 is $1\frac{98}{100}$, etc. Between 2 and 3 the divisions are tenths and half-tenths, a half-tenth being $\frac{5}{100}$. From 3 to 10 the divisions are all tenths; from 10 to 20 each subdivision represents $\frac{2}{10}$; the first division beyond 20 represents $20\frac{5}{10}$; the second division represents 21; the third division represents $21\frac{5}{10}$; the fourth represents 22, and so on up to 30; from 30 up to 100 each division represents 1.

These numbers, marked 1, 2, 3, etc., are arbitrary, and have no fixed values; for, beginning at the left, 1 might represent 10; 2 would represent 20; each of the principal divisions between 1 and 2, which in the ordinary reading represented tenths, would represent 1; each of the subdivisions, which in the ordinary reading represented $\frac{2}{100}$, would represent $\frac{2}{10}$; 3 would represent 30; the number which formerly represented 10 would represent 100; the number which formerly represented 12 would represent 120; the number formerly representing 20 would represent 200, and so on, the value of the whole line being increased tenfold; or, 1 at the left might represent 100, 2 would represent 200, and so on, the value of the whole line being increased one hundred-fold. On different lines, 1 may bear different values in working out a problem. For example: Multiply 40 by 5. We set 1, which is on the line B, to 5, on the line A: above 4, which we will call 40, on the line B, we find 20 on the line A; but, since we have increased the value of one of the divisions tenfold its ordinary value, we must increase the result the same, which gives us 200 as the answer. The line D is divided the same as A, B, and C are, from 1 to 10, only on a

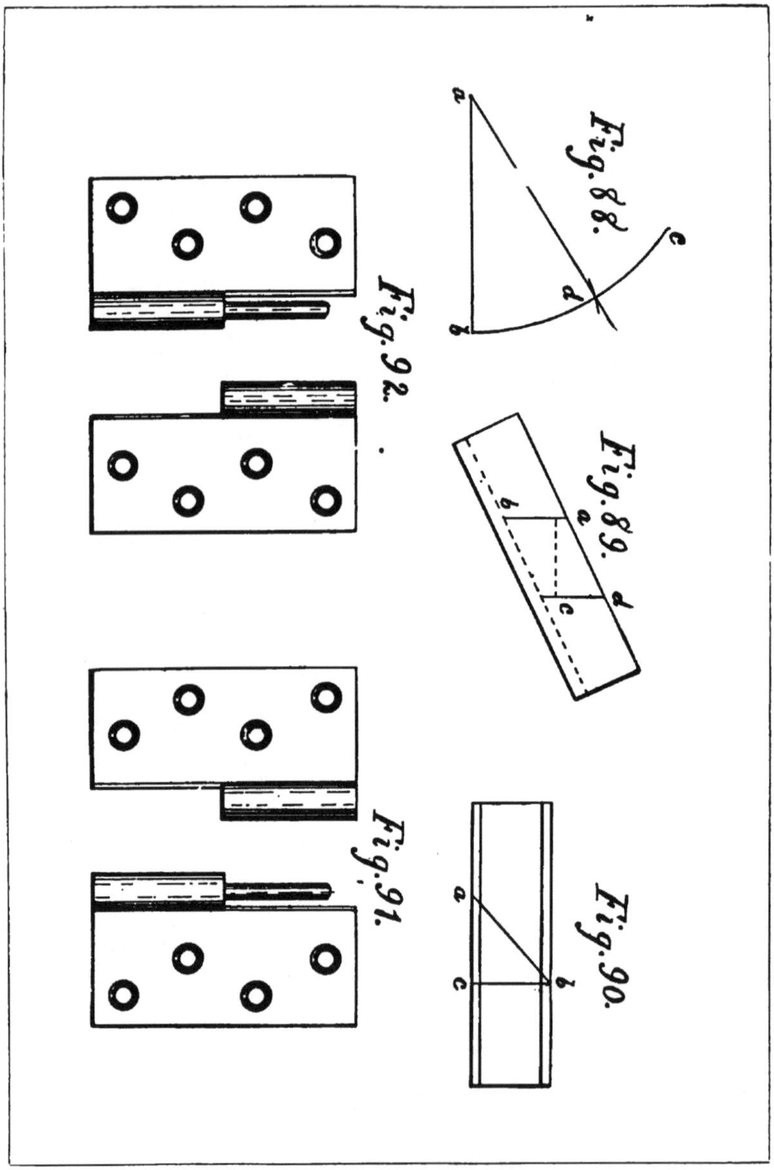
Fig. 88.
a
b
c
d
Fig. 89.
a
b
c
d
Fig. 90.
a
b
c
Fig. 91.
Fig. 92.

larger scale; and 1 on this line may represent 1, 10, or 100, the same as the other lines. It will require considerable practice to readily and correctly read the numbers and tenths or hundredths on the slide-rule, with the different values which 1 may bear; and, in practising, it would be well for the beginner to compare the answers he obtains with some printed tables that are correct. If his answers do not agree with the tables, he has made an error somewhere, which must be rectified. By considerable and careful practice he will become expert in the use of the slide-rule.

Multiplication by the Slide-rule. — RULE. Set 1 on the line B to the number on A, which is used as the multiplier: then above the number on B, which is used as a multiplicand, find the answer on the line A.

Examples. — To multiply 4 by 5, we set 1 on the line B to 4 on the line A: then above 5 on the line B we find the answer 20 on the line A.

To multiply $3\frac{1}{2}$ by $2\frac{1}{2}$, we set 1 on the line B to $2\frac{1}{2}$ on the line A: then above $3\frac{1}{2}$ on the line B we find the answer $8\frac{3}{4}$ on the line A.

To multiply 30 by 4, we set 1 on the line B to 4 on the line A: then above 3, which we will call 30, on the line B, we find 12 on the line A. Now, as we have increased the value of three tenfold over its ordinary value, we must increase the result tenfold to get the answer: 10 times 12 equal 120, the required answer.

To multiply 35 by 25, we set 1 on the line B to $2\frac{1}{2}$ (2.5, or $2\frac{5}{10}$), which we will call 25, on the line A: then above $3\frac{1}{2}$ (3.5, or $3\frac{5}{10}$), which we will call 35, on the line B, we find 8.75 ($8\frac{7\frac{1}{2}}{10} = 8\frac{75}{100}$) on the line A. Now, as we have increased the value of $2\frac{1}{2}$ tenfold, and also have increased the value of $3\frac{1}{2}$ to tenfold its ordinary value, we must increase the result ten times tenfold, which is one hundred-fold: one hundred times 8.75 ($8\frac{75}{100}$) is 875, the required answer.

Division by the Slide-rule. — RULE. Set the number indicating the divisor on the line B under the number indicating the dividend on the line A: then above 1 on the line B find the answer on the line A.

Examples. — To divide 24 by 6, we set 6 on the line B under 24 on the line A: then above 1 on the line B we find the answer 4 on the line A.

To divide 260 by 13, we set 13 on the line B under 26, which we will call 260, on the line A: then over 1 on the line B we find 2 on the line A. But, since we have increased the value of 26 tenfold its ordinary value, we must increase the result tenfold: ten times 2 equal 20, the required answer.

To divide 3,500 by 50, we set 5, which we will call 50, on the line B under 35, which we will call 3,500, on the line A: then above 1 on the line B we find 7 on the line A. Now, to find how many fold to increase this result, we divide the number of times we increased the value of 35, which we increased one hundred-fold, by the number of times we increased the value of 5, which was tenfold; 100 divided by 10 equals 10, so we must increase the result tenfold; ten times 7 equal 70, the required answer.

Proportion by the Slide-rule. — Example 1. As 3 is to

; we
lave
. we
one
, the

nber
nber
1 on

ne B
B we

r 26.
n the
e in-
, we
20,

l 50,
the
line
sult,
e of
nber
old;
the
ver.

is to

HOUSE-CARPENTERS

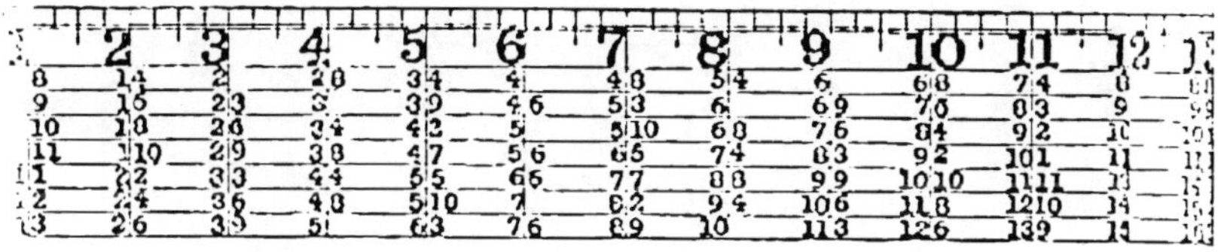

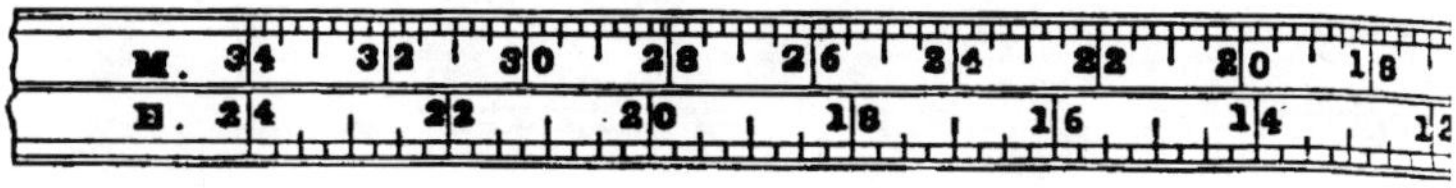

PLATE 37.

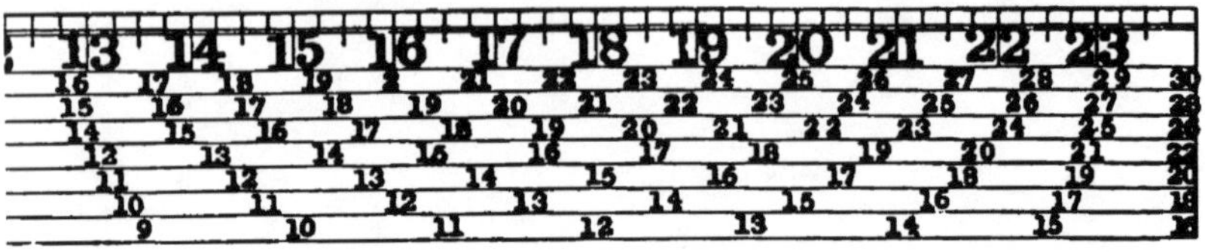

93

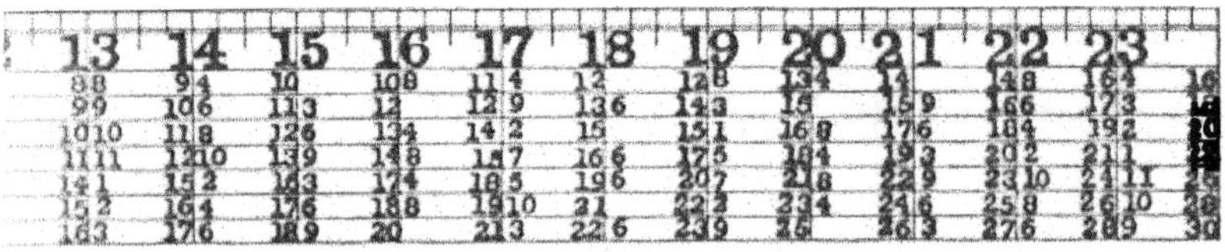

94

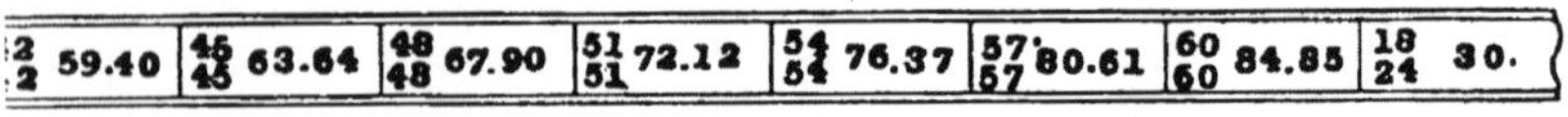

95

96

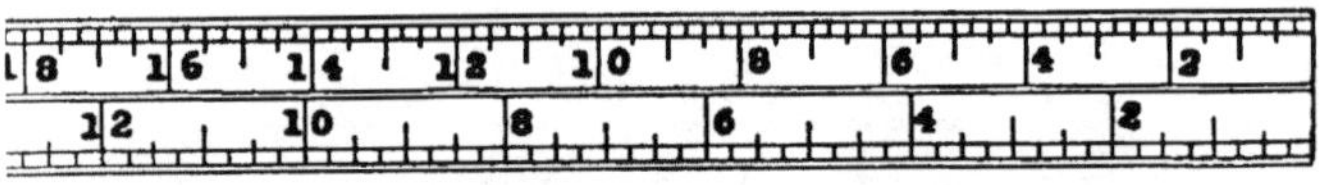

97

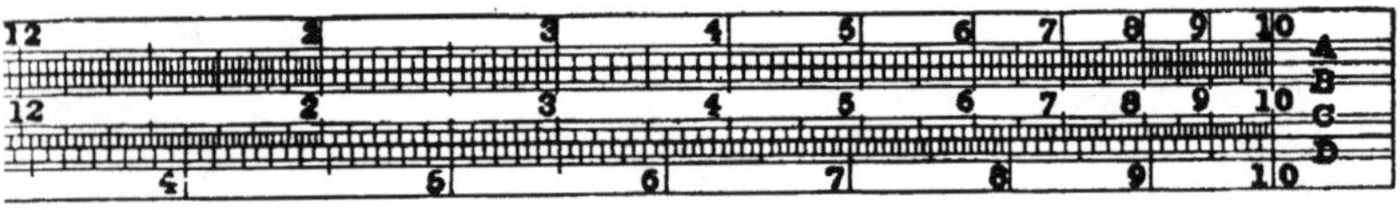

10

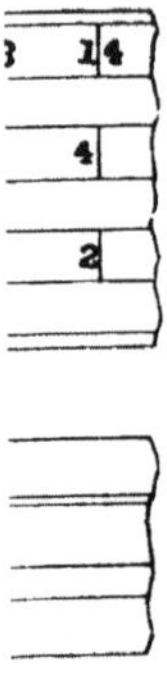

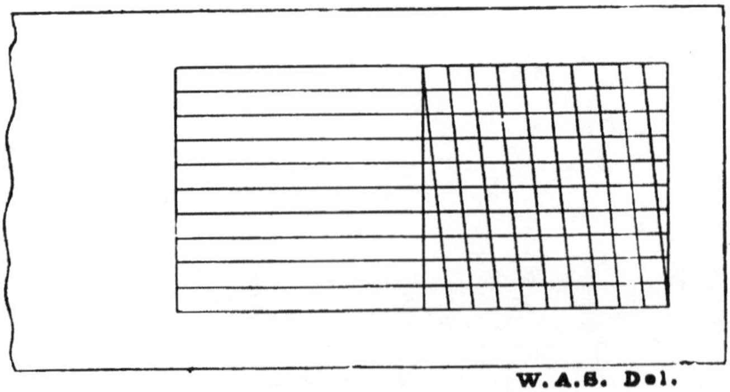

100

W. A. S. Del.

IGHT.

12, so is 5 to the answer. We set 3 on the line B under 12 on the line A: then above 5 on the line B we find the answer, 20, on the line A.

Example 2. — As $2\frac{1}{2}$ is to $5\frac{1}{4}$, so is 3 to the answer. We set $2\frac{1}{2}$ on the line B under $5\frac{1}{4}$ on the line A: then above 3 on the line B we find the answer, $6\frac{3}{10}$, on the line A.

Example 3. — If my wages are $2.75 per day (working 10 hours), how much will be due me when I have worked 6 days and 4 hours? We reduce the days and parts of a day to hours: 6 days and 4 hours equal 64 hours: so we state our proportion as follows. As 10 is to 64, so is $2.75 to the answer. We set 1, which we will call 10, on the line B, under $6\frac{4}{10}$ (6.4), which we will call 64, on the line A: then above $2\frac{7\frac{1}{2}}{10}$ (2.75, or $2\frac{3}{4}$) on the line B we find the answer, $17\frac{3}{5}$, on the line A. \$$17\frac{3}{5}$ equal $17.60.

Example 4. — If I pay a man $15 per week, what should I pay him for $8\frac{1}{2}$ days' work? We may state our proportion as follows. As 6 is to $8\frac{1}{2}$, so is 15 to the answer. We set 6 on the line B under $8\frac{1}{2}$ on the line A: then above 15 on the line B we find the answer, $21\frac{1}{4}$, on the line A. \$$21\frac{1}{4}$ equal $21.25.

SQUARES AND SQUARE ROOTS.

The square of any number is the result obtained by multiplying that number by itself.

The square root of any number is that number which, when multiplied by itself, will produce the given number.

When the slide is shut so that 1 on the line C is even with 1 on the line D, then the square of any number on the line D is found just above it on the line C: thus the square of 3 on the line D is 9, which is found just above it on the line C; and the square root of any number on the line C is found just below it on the line D: thus the square root of 9 on the line C is 3, which is found just below it on the line D. The square root of 5 on the line C is 2.23 ($2\frac{23}{100}$), which is found just below it on the line D.

DECIMAL AND COMMON FRACTIONS.

To change a Common Fraction to a Decimal. — RULE. Set the number representing the denominator of the given fraction on the line B, under the number representing the numerator: then above 1 on the line B the number of tenths or hundredths will be found on the line A.*

Example. — Change $\frac{3}{4}$ to a decimal. We set the denominator 4 on the line B, under the numerator 3 on the line A (actually under 30): then above 1 on the line B we find $7\frac{1}{2}$ tenths, or 75 hundredths (.75), on the line A.

To change a Decimal to a Common Fraction. — RULE. Set 1 on the line B, under the number of tenths or hundredths on the line A; then find the number on the line A which exactly coincides with any number on the line B; the number on the line A will be the numerator, and the

* The right-hand half of the line A is generally used in stating the example; what really is 10 being considered as 1, and 20 being considered as 2, etc., unless the numerator of the given fraction should be larger than 10, in which case the numerator would be the number actually taken on the line A. In many cases, where the numerator and denominator contain more than one figure, considerable judgment must be used to determine whether the result obtained is tenths, hundredths, or thousandths. From 1 to 10 on the line A is divided into 10 parts, which are numbered from 1 to 10. In changing fractions to decimals, these parts usually represent tenths. These parts are subdivided into 10 parts each, which represent hundredths; there being 100 divisions, all told, from 1 to 10.

number on the line B will be the denominator, of the required fraction.

Example. — Change the decimal .625 to a common fraction. We set 1 on the line B, to $6\frac{1}{4}$ ($6\frac{1}{4}$ tenths = .625) on the line A; then we find, toward the right-hand end of the rule, that 5 on the line A coincides with 8 on the line B: so we have $\frac{5}{8}$ for the answer.

MISCELLANEOUS.

To find the Diagonal of Any Square (or the Length of Brace where the Run is the Same Each Way). — Set 70 on B to 99 on A: find the diagonal on A above the length of side (or run, for braces) on B.

To find the Diameters or Circumferences of Circles. — Set 7 on B to 22 on A: then any number on A is the circumference of the circle whose diameter is immediately beneath, or *vice versa*.

BOARD MEASURE.

Set the length in feet on B to 12 on A: find the number of square feet on B below the width in inches on A.

AREAS OF CIRCLES.

Set 7 on C to 3 on D: the area of any circle whose diameter is on D is found just above, on C.*

We have given only a small part of what may be done by the use of the slide-rule. We have given only that which would be of the most value to workmen. There

* If the diameter is inches, the area will be square inches; and, if the diameter is feet, the area will be square feet.

are several works which are devoted entirely to explanations of the uses of the slide-rule, and any one desiring to pursue this subject farther would do well to obtain a copy. However, until one is very expert in the use of the slide-rule, it would be safer for him to figure out the problems he may have occasion to do.

GLUE AND GLUING.

There are many varieties of glue, ranging in price from twelve cents to fifty cents per pound. For general use, a good quality of glue can be purchased for twenty or twenty-five cents per pound. Previous to cooking, glue should be soaked in cold water till it becomes quite soft and pliable; the length of time required depends on the kind and quality of the glue: poor, cheap glue will nearly, and sometimes completely, dissolve in cold water; while good glue will require several hours' soaking; some kinds require to be soaked twenty-four hours or more, but such glue is not commonly used. When the glue has been soaked sufficiently, drain off what water remains, and set the dish holding the glue into a dish containing water, and set it over the fire to cook. The object of setting the glue-dish into water, is to prevent the glue from getting scorched. The water cannot get hotter than 212°, which is not hot enough to injure the glue. To secure the best possible results, the following conditions must be complied with: namely, the glue must be of good quality and newly made; it must be of the proper consistency, neither too thick — or the two surfaces will not come together — nor yet too thin; the glue must be as hot as boiling water can heat it; the work must be properly fitted, and should be as warm as can be borne against the cheek; the room should be very warm, especially in gluing large surfaces, and in veneering: the glue should be plentifully applied to both surfaces, and then the work should be clamped together firmly; and the clamps should not be taken off until the glue is hard, clear into the middle of the joints. Very large jobs of gluing should set two or three days before the clamps are removed. The consistency of the glue will depend somewhat on the kind of work to be done. For large surfaces, the glue may be quite thin, and plentifully used. For small work, the glue may be of thicker consistency; but it must be applied hot. For gluing wood endways, the ends should first be sized with a *very* thin coat of glue; when the sizing gets thoroughly dry, smooth the raised grain with a piece of fine sand-paper used over a straight stick; then coat each end with hot glue, and clamp firmly together; let it set over night, sure. In gluing boards together edgeways, many workmen do not bother to joint them both true, but depend on the clamps to force them to a joint. If the glue is good, the work may hold together some time; but there is always a strain on the glue. Some spell of damp weather may soften the glue a *very little*, and open goes the joint. Of course, the glue gets the blame, instead of the workman, who deserves to be blamed. While many workmen make rubbed joints six feet or more in length, it is a bad practice: no joint longer than two feet ought ever to be merely rubbed together, and it is safer to apply clamps in every case. In veneering, put a thickness of newspaper between the veneer and the caul. This prevents the glue, which strikes through the veneer, from sticking the veneer to the caul. Some accomplish the same purpose by using sheets of zinc, which they rub with a piece of hard soap or wax. This is better than using paper, as it saves the labor of cleaning the paper off from the veneer.

To keep glue from smelling, take the dish holding the glue out of the dish containing the water, when done using, so as to let the glue get cold as soon as possible. Also do not keep the glue boiling all day long, but heat it only when it is needed. It is a good plan to make up only enough at a time to last two or three days, especially in the summer-time, so as to have it fresh and good. A piece of sheet-zinc, as large as will lay in the bottom of the glue-pot, will also greatly aid in keeping the glue from smelling. Some put in a little alcohol, but it is doubtful whether it does any good: it probably very quickly evaporates. Oil of cloves would be better.

N.B. — Keep thin-shaved veneers, such as ash and walnut burls, in rather a damp place until wanted; as they will curl and split up badly if kept in a dry room.

STRENGTH OF MATERIALS.

TENSILE STRENGTH OF MATERIALS.

WEIGHT OR POWER REQUIRED TO TEAR ASUNDER ONE SQUARE INCH.

METALS.	Breaking-Weight.	WOODS.	Breaking-Weight.
	lbs.		lbs.
Copper, wrought	34,000	Ash	14,000
" cast (Amer.)	24,250	Beech	11,500
" bolt	36,800	Box	20,000
Iron, cast	31,829	Bay	14,000
" " (safe-load)	4,000 to 5,000	Cedar	11,400
		Chestnut, sweet	10,500
		Cypress	6,000
" wrought	55,000 to 65,000	Deal, Christiana	12,400
		Elm	13,400
		Lance	23,000
" " (safe-l'd),	8,000 to 10,000	Lignum-vitæ	11,800
		Locust	20,500
		Mahogany	21,000
" bolts	52,250	" Spanish	12,000
Lead, cast	1,800	" "	8,000
Steel, mean	88,657	Maple	10,500
" maximum	142,000	Oak, American white,	11,500
Tin, cast block	5,000	" English	10,000
" Banca	2,122	" " (seasoned)	13,600
Yellow metal	48,700	" African	14,500
Zinc	3,500	Pear	9,800
" sheet	16,000	Pine, pitch	12,000
		" larch	9,500
		" American white,	11,800
		Poplar	7,000
		Spruce, white	10,290
		Sycamore	13,000
		Teak	14,000
		Walnut	7,800
		Willow	13,000

RESULTS OF EXPERIMENTS ON THE TENSILE STRENGTH OF WROUGHT-IRON TIE-RODS.*

COMMON ENGLISH IRON, $1\frac{3}{16}$ INCHES IN DIAMETER.

DESCRIPTION OF CONNECTION.	Breaking-Weight.
	lbs.
Semicircular hook fitted to a circular and welded eye . .	14,000
Two semicircular hooks hooked together	16,220
Right-angled hook, or gooseneck, fitted into a cylindrical eye	29,120
Two links, or welded eyes, connected together	48,160
Straight rod without any connection articulation . . .	56,000

TRANSVERSE STRENGTH OF MATERIALS.

When one end is fixed, and the other projecting, the strength is inversely as the distance of the weight from the section acted upon; and the strain upon any section is directly as the distance of the weight from that section.

When both ends are supported only, the strength is 4 times greater for an equal length, when the weight is applied in the middle between the supports, than if one end only is fixed.

When both ends are fixed, the strength is 6 times greater for an equal length, when the weight is applied in the middle, than if one end only is fixed, or one-half stronger than if both ends were merely supported.†

When the weight or strain is uniformly distributed, a beam will sustain double the weight that it would bear if the load was all at the centre.

* From one-fourth to one-seventh of the breaking-weight is a safe-load.

† If a beam is supported two or three feet from each end, a weight applied in the centre would cause the ends to tip up as the middle went down; but if the ends were *fixed*,—say, for example, built into a brick wall,—the beam would sustain one-half more weight than if the ends were merely supported.

TRANSVERSE STRENGTH OF MATERIALS.*

REDUCED TO THE UNIFORM MEASURE OF ONE INCH SQUARE, AND ONE FOOT IN LENGTH, EXTENDING HORIZONTALLY, FIXED AT ONE END, WEIGHT SUSPENDED FROM THE OTHER.

METALS.	Breaking-Weight.	Safe-Load.
	lbs.	lbs.
Cast-iron	507 to 772	125 to 250
" " mean . .	681	170 to 225
Wrought-iron . . .	600 to 700	160 to 200
Steel (greatest) . .	1,918	350 to 450
" puddled (permanent bent)	800	170 to 225
Copper	—	55
Brass	—	58
STONES (American).		
Flagging (blue) . .	31	10
Freestone (Conn.) .	13	4
" (Dorchester	10.8	3.5
Freestone (N.Jersey)	17.8 to 20.1	6 to 6.5
Freestone (N.York),	24	8
Granite, blue, coarse,	18	6
" (Quincy, Mass.)	26	8.5

WOODS.	Breaking-Weight.	Safe-Load.
	lbs.	lbs.
Ash	168	55
Beech	130	32
Birch	160	40
Chestnut	160	53
Deal (Christiana) . .	137	45
Elm	125	30
Hickory.	250	65
Locust	295	80
Maple	202	65
Norway Pine	123	40
Oak, African	208	50
" American white .	230	50
" " live .	245	55
" English	140 to 188	35 to 45
Pitch-pine (American)	160	50
Riga Fir	94	30
Teak	206	60
White Pine (Amer.) .	130	45
Whitewood	116	38

* The safe-load of any material is from one-fourth to one-seventh of its breaking-weight.

TO COMPUTE THE TRAVERSE STRENGTH OF A RECTANGULAR BEAM OR BAR.

When the Beam or Bar is fixed at One End, and loaded at the Other. — RULE. Multiply the safe-load given in the table by the breadth and the square of the depth in inches, and divide the product by the length in feet.*

If the Dimensions are required of a Beam or Bar, supported at one End to sustain a Given Weight at the Other End. — RULE. Divide the product of the weight and the length in feet by the safe-load given in the table, and the result is the square of the depth multiplied by the breadth or thickness: so by dividing this result by the breadth, and extracting the square root of the quotient, we have the depth in inches.

When a Beam or Bar is fixed at Both Ends, and loaded in the Middle. — RULE. Multiply the safe-load given in the table by 6 times the breadth, and by the square of the depth in inches, and divide the product by the length in feet.

If the Dimensions of a Beam or Bar are required to support a Given Weight in the Middle, between the Fixed Ends. — RULE. Divide the product of the weight and the length in feet by 6 times the safe-load given in the table, and the quotient will be the square of the depth multiplied by the breadth or thickness in inches: so we divide this result by the breadth, and extract the square root of the quotient, which gives the depth; or, divide

* When the beam is loaded uniformly throughout its length, the result must be doubled.

the result by the square of the depth, and the quotient is the breadth or thickness.

When a Beam or Bar is supported at Both Ends, and loaded in the Middle. — RULE. Multiply the safe-load given in the table by 4 times the breadth, and by the square of the depth in inches, and divide this product by the length in feet.*

If the Dimensions are required to support a Given Weight. — RULE. Divide the product of the weight and the length in feet by 4 times the safe-load given in the table; the result is the square of the depth multiplied by the breadth or thickness: so we divide this result by the breadth, and extract the square root, which gives the depth; or, divide the result by the square of the depth, and the quotient is the breadth or thickness in inches.

In all uses, such as in buildings and bridges, where the structure is exposed to sudden impulses, the load or stress to be sustained should not exceed from $\frac{1}{5}$ to $\frac{1}{6}$ of the breaking-weight of the material employed; but when the load is uniform, or the stress quiescent, it may be increased to $\frac{1}{3}$ or $\frac{1}{4}$ of the breaking-weight. In churches, buildings, etc., the weight to be provided for should be estimated at that which at any time may be placed thereon, or which at any time may bear upon any portion of their floors. Where the weight of people alone is to be provided for, an estimate of 175 pounds per square foot of floor-surface is sufficient to provide for the weight of flooring and the

* When the beam is loaded uniformly throughout its length, the result must be doubled.

load upon it. The usual allowance for stores and factories is 280 pounds per square foot of floor-surface.

When a beam has four or more supports, its condition as regards a stress upon its middle is that of a beam fixed at both ends.

I WROUGHT-IRON BEAMS.*

(Trenton Iron Works, Cooper, Hewitt, & Co., New York.)

Depth.	Thickness of Web.	Width of Flanges.	Weight per Foot in Length.	Safe-Load.	Depth.	Thickness of Web.	Width of Flanges.	Weight per Foot in Length.	Safe-Load.
in.	in.	in.	lbs.	lbs.	in.	in.	in.	lbs.	lbs.
6	$\frac{1}{4}$	3	13.3	76,000	9	$\frac{1}{2}$	4	30	246,000
6	$\frac{5}{16}$	$3\frac{1}{4}$	16.6	92,000	9	$\frac{5}{8}$	$5\frac{3}{8}$	50	448,000
7	$\frac{3}{8}$	$3\frac{1}{2}$	20	124,000	$12\frac{1}{4}$	$\frac{7}{16}$	$4\frac{1}{2}$	40	390,000
9	$\frac{3}{8}$	$3\frac{1}{2}$	23.3	192,000	15	$\frac{9}{16}$	$4\frac{1}{16}$	51.6	640,000
9	$\frac{7}{16}$	4	28	240,000	15	$\frac{5}{8}$	$5\frac{3}{8}$	66.6	908,000

To find the Safe-load for any of the Above Beams for a Given Length, Weight to be uniformly distributed. — Rule. Divide the safe-load of the beam given in the table by the length in feet.

Illustration. — What is the weight, uniformly distributed, that may be borne with safety by an iron beam 6 inches deep, web $\frac{5}{16}$ thick, flanges $3\frac{1}{4}$ inches wide, and 10 feet long? We find the safe-load given in the above table to be 92,000 pounds, which, divided by ten, the length in feet, gives 9,200 pounds.

* Load uniformly distributed, beam resting upon two supports.

CRUSHING-STRENGTH OF MATERIALS.

Reduced to a Uniform Measure of One Square Inch.

MATERIAL.	Crushing-Weight.	MATERIAL.	Crushing-Weight.
METALS.	lbs.	STONES, ETC.	lbs.
Cast-iron, American, mean	129,000	Common brick masonry	500 to 800
Wrought-iron, American	127,720		
" " mean,	83,500	Freestone, Belleville	3,522
Fine brass	164,800	" Caen	1,088
Cast-copper	117,000	" Connecticut	3,919
Cast-steel	295,000	" Dorchester	3,069
Cast-tin	15,500	" Little Falls	2,991
Lead	7,730	Granite, Patapsco	5,340
		" Quincy	15,300
		Marble, Baltimore, large	8,057
WOODS.		" " small	18,061
		" East Chester *	13,917
Ash	6,663	" Hastings, N.Y.	18,941
Beech	6,963	" Italian	12,624
Birch	7,969	" Lee, Mass.	22,702
Box	10,513	" Montgomery Co., Penn.	8,950
Cedar, red	5,968		
Chestnut	5,350	" Stockbridge †	10,382
Elm	6,831	" Symington, large	11,156
Hickory, white	8,925	" " fine crystal	18,248
Locust	9,113		
Mahogany, Spanish	8,198	" Symington, strata horizontal	10,124
Maple	8,150		
Oak, American white	6,100	" Symington, strata vertical	9,324
Pine, pitch	8,947		
" white	5,775	Mortar	120 to 240
" yellow	8,200		
Spruce, white	5,350		
Sycamore	7,082	Sandstone, Adelaide	2,800
Teak	12,100	" Aquia Creek ‡	5,340
Walnut	6,645	" Seneca §	10,762

* Same as that of the General Post-office, Washington.

† Same as that of the City Hall, New York.

‡ Same as that of the Capitol, Treasury Department, and Patent Office, Washington, D.C.

§ Same as that of the Smithsonian Institute.

The Crushing-strength of any body is in proportion to the area of its section, and inversely as its height.

In tapered columns the strength is determined by the least diameter.

With cast-iron, a pressure beyond 26,680 pounds per square inch is of little, if any, use in practice.

The safe-load that may be borne by a column of cast-iron, independent of any considerations, regarding the operation of its ends, as to their being flat or rounded, etc., is from 5,000 to 8,000 pounds per square inch for short or stable bodies.

THE NUMBER OF 1,000 LBS. THAT CAN BE BORNE WITH SAFETY BY SOLID CAST-IRON COLUMNS. (TRENTON IRON COMPANY.)

Length.	2 in.	3 in.	4 in.	5 in.	6 in.	7 in.	8 in.	9 in.	10 in.	11 in.	12 in.	13 in.	14 in.	15 in.
5 ft.	12.4	44	102	184	288	414	560	728	916	1,126	1,354	–	–	–
6 "	9.4	36	88	164	264	386	532	698	884	1,082	1,320	1,570	–	–
7 "	7.2	30	76	146	242	360	502	660	850	1,056	1,282	1,530	1,798	2,086
8 "	–	24	66	130	218	332	470	630	812	1,016	1,240	1,486	1,754	2,040
9 "	–	20	56	114	198	306	440	596	774	974	1,196	1,440	1,706	1,992
10 "	–	18	48	102	180	282	410	560	739	932	1,152	1,392	1,656	1,940
12 "	–	–	38	80	136	238	354	494	658	846	1,056	1,292	1,550	1,828
14 "	–	–	28	64	122	200	304	432	586	774	966	1,192	1,440	1,712
16 "	–	–	–	52	100	170	262	378	520	686	878	1,094	1,332	1,596
18 "	–	–	–	44	84	144	226	332	462	616	796	1,000	1,228	1,482
20 "	–	–	–	–	72	124	196	292	410	552	720	912	1,130	1,372

For tubes or hollow columns, subtract the weight that may be borne by a column of the size of the internal diameter of the tube or column. The thickness of metal should not be less than one-twelfth the diameter.

MATHEMATICAL RULES, ETC.,

FOR THE CONVENIENCE OF MECHANICS.

Indicative Characters or Signs.

The sign + (plus) between two numbers indicates that they are to be added together.

The sign − (minus) indicates that the number placed after it is to be subtracted from the number placed before it.

The sign × (times) indicates that one number is to be multiplied by another.

The sign ÷ (divided by) indicates that the number on the left hand is to be divided by the number on the right hand.

The sign = (equal to) indicates that the result of the figures before it amounts to the number placed after it.

The sign √ is called the radical sign; and, if it has a figure 2 placed over it, it signifies that the *square root* of the number before which it is placed is required. If it has the figure 3 placed over it, then it is the *cube root* which is required.

In figuring drawings, feet are usually indicated by a single index, and inches are indicated by two indices, thus: $12' \, 8\frac{3}{4}''$ is 12 feet $8\frac{3}{4}$ inches.

Prime Numbers.

A prime number is a number that cannot be divided by any other number without leaving a remainder.

TABLE OF PRIME NUMBERS FROM 1 TO 1,000.									
1	59	139	233	337	439	557	653	769	883
2	61	149	239	347	443	563	659	773	887
3	67	151	241	349	449	569	661	787	907
5	71	157	251	353	457	571	673	797	911
7	73	163	257	359	461	577	677	809	919
11	79	167	263	367	463	587	683	811	929
13	83	173	269	373	467	593	691	821	937
17	89	179	271	379	479	599	701	823	941
19	97	181	277	383	487	601	709	827	947
23	101	191	281	389	491	607	719	829	953
29	103	193	283	397	499	613	727	839	967
31	107	197	293	401	503	617	733	853	971
37	109	199	307	409	509	619	739	857	977
41	113	211	311	419	521	631	743	859	983
43	127	223	313	421	523	641	751	863	991
47	131	227	317	431	541	643	757	877	997
53	137	229	331	433	547	647	761	881	

Long-Measure Table.

12 inches make	1 foot
3 feet make	1 yard
$5\frac{1}{2}$ yards, or $16\frac{1}{2}$ feet, make	1 rod or pole
40 rods make	1 furlong
8 furlongs make	1 mile

One mile contains 5,280 feet, or 1,760 yards, or 320 rods; 3 miles make 1 league; 6 feet=1 fathom.

Surface, or Square Measure.

144 square inches make	1 square foot
9 square feet make	1 square yard
30¼ square yards make	1 square rod
160 square rods make	1 acre
640 acres make	1 square mile

Lathing and plastering are usually reckoned by the square yard. Of flooring, slating, etc., a square is 100 square feet.

Cubic, or Solid Measure.

1728 cubic inches make	1 cubic foot
27 cubic feet make	1 cubic yard
16 cubic feet make	1 cord foot
8 cord feet, or 128 cubic feet, make .	1 cord of wood

A pile of wood 8 feet long, 4 feet broad, and 4 feet high contains a cord.

A cord foot is 1 foot in length of the above pile.

A perch of masonry is 16½ feet long, 1 foot high, and 18 inches thick; or 24¾ cubic feet.

To reduce Several Fractions to their Least Common Denominator.

The numerator of a fraction is the number above the line; the denominator is the number below the line.

Rule. — 1. Find the least common multiple of the denominators for a new denominator.

2. Divide the least common denominator by each given denominator, and multiply the quotient by the corresponding numerator, for the new numerators.

Example.—Reduce $\frac{1}{2}$, $\frac{3}{4}$, $\frac{1}{3}$, $\frac{5}{6}$, and $\frac{1}{8}$ to the least common denominator.

2	2 — 4 — 3 — 6 — 8
2	2 — 3 — 3 — 4
3	3 — 3 — 2
2	2

$2 \times 2 \times 3 \times 2 = 24$, the least common denominator.

$$\frac{1}{2})24 \quad 12 \times 1 = 12 \qquad \frac{1}{2} = \frac{12}{24}$$

$$\frac{3}{4})24 \quad 6 \times 3 = 18 \qquad \frac{3}{4} = \frac{18}{24}$$

$$\frac{1}{3})24 \quad 8 \times 1 = 8 \qquad \frac{1}{3} = \frac{8}{24}$$

$$\frac{5}{6})24 \quad 4 \times 5 = 20 \qquad \frac{5}{6} = \frac{20}{24}$$

$$\frac{1}{8})24 \quad 3 \times 1 = 3 \qquad \frac{1}{8} = \frac{3}{24}$$

Explanation.—We first find the least common multiple of the denominators, 2, 4, 3, 6, and 8, by dividing them by a number which is contained in one or more of them; and this quotient we divide again in the same manner, and so on until the division is complete; then multiply together the divisors, and the result is the least common multiple of these denominators, which we use for the new denominator. Then divide this new denominator by each of the others, and multiply this quotient by the given numerators.

To reduce a Fraction to a Given Denominator.

It sometimes happens in figuring out work that the fractions come different from any marking of the rule. For instance, it may come in ninths, or in fourteenths; the workman wants to know how many eighths and sixteenths a certain number of ninths or fourteenths may be. Suppose we have $\frac{11}{14}$ of an inch, we want to know how many sixteenths that represents.

Rule.—Multiply the required denominator by the numerator, and divide the product by the denominator of the given fraction: the result will be the required numerator.

Thus 16, the required denominator, multiplied by 11, the numerator, gives 176; which, divided by 14, the denominator of the given fraction, gives $12\frac{8}{14}$, the new numerator: so that $\frac{11}{14} = \frac{12\frac{8}{14}}{16} = \frac{3}{4}$ strong.

To reduce Fractions to Decimals.

Rule. — 1. Annex ciphers to the numerator, and divide by the denominator.

2. Point off in the quotient as many decimal places as there have been ciphers annexed.

EXAMPLE. — Reduce $\frac{1}{8}$ to a decimal.

Ans. .125 (125 thousandths).

$$8)\underline{1.000}$$
$$.125$$

Simple Proportion, or Rule of Three.

Simple Proportion is an equality between two simple ratios.

Ratio is the relation, in respect to magnitude or value, which one quantity or number has to another of the same kind; or the quotient arising from the division of one number by another: thus, the ratio of 8 to 4 is 2, since 8 is 2 times 4; the ratio of 4 to 8 is $\frac{1}{2}$, since 4 is $\frac{1}{2}$ of 8.

Rule. — Make that number the third term which is of the same kind as the answer; and if, from the nature of the question, the third term must be greater than the fourth term, or answer, make the greater of the two remaining terms the first term, and the smaller, the second; but, if the third term must be less than the fourth, make the less of the two remaining terms the first, and the greater, the second; then multiply the second and third terms together, and divide their product by the first term: the quotient will be the fourth term, or answer.

EXAMPLES. — If a man receives $15 for a week's work, how much shall he have for 7 days' work?

da. da. $ $
6 : 7 : : 15 : (   )
7
6)105
$$17\frac{3}{6} = \$17.50.$$ *Ans.*

If 5 men can build a house in 45 days, how long will it take 8 men?

m. m. da. da.
8 : 5 : : 45 : (   )
5
8)225
$28\frac{1}{8}$ days. *Ans.*

Compound Proportion.

Compound Proportion is an expression of equality between a compound and a simple ratio.

Rule. — Make that number the third term which is of the same kind as the answer; of the remaining numbers, take any two that are of the same kind, and consider whether an answer depending upon these alone would be greater or less than the third term, and place them as directed in simple proportion.

Then take any other two of the same kind, and consider whether an answer depending only upon them would be greater or less than the third term, and arrange them accordingly; and so on until all are used. Multiply the product of the second terms by the third term, and divide the result by the product of the first terms: the quotient will be the fourth term, or answer.

Example. — If 6 men can build an 8-inch brick wall, 95 feet long and 15 feet high, in 3 days, how long will it take 5 men to build a 12-inch wall, 40 feet long and 9 feet high, the days being 10 hours long in both cases?

5 men	: 6 men	
8 in.	: 12 in.	: : 3 da. : (da.)
95 long	: 40 long	
15 high	: 9 high	

$5 \times 8 \times 95 \times 15 = 57000$

$6 \times 12 \times 40 \times 9 \times 3 = 77760 \div 57000 = 1.36+$ days $= 1$ day 3 hours 36 + minutes.

Example. — I paid $35 for the labor of 2 men for 6 days, they working 12 hours daily. How much ought I to pay 4 men for 7 days' work, 10 hours being reckoned a day's work, and paying at the same rate per hour as I paid the first men?

2 men	: 4 men	
6 da.	: 7 da.	: : $35 : ($)
12 h.	: 10 h.	

$2 \times 6 \times 12 = 144$

$4 \times 7 \times 10 \times 35 = 9800 \div 144 = 68.05$. *Ans.* — $68.05.

Square Root.

The Square Root of any number is that number which, multiplied by itself, will produce the given number.

Rule for extracting the Square Root. — 1. Point off the given number into periods of two figures each; counting from units' place toward the left in whole numbers, and toward the right in decimals.

2. Find the greatest square number in the left-hand period, and write its root for the first figure in the root; subtract the square number from the left-hand period, and to the remainder bring down the next period for a dividend.

3. At the left of the dividend write twice the first figure of the root, and annex one cipher for a trial divisor; divide the dividend by the trial divisor, and write the quotient for a trial figure in the root.

4. Add the trial figure of the root to the trial divisor for a complete divisor; multiply the complete divisor by the trial

figure in the root, and subtract the product from the dividend; and to the remainder bring down the next period for a new dividend.

5. To the last complete divisor add the last figure of the root, and to the sum annex one cipher for a new trial divisor, with which proceed as before.

Note 1. — If at any time the product be greater than the dividend, diminish the trial figure of the root, and correct the erroneous work.

Note 2. — The left-hand period may contain but one figure.

Note 3. — If the dividend does not contain the divisor, a cipher must be placed in the root, and also at the right of the divisor; then, after bringing down the next period, this last divisor must be used as the divisor of the new dividend.

Note 4. — When there is a remainder after extracting the root of a number, periods of ciphers may be annexed; and the figures of the root thus obtained will be decimals.

Note 5. — The square root of a fraction may be obtained by extracting the square roots of the numerator and denominator separately, providing the terms are perfect squares; otherwise the fractions must first be reduced to decimals.

Examples. — What is the square root of 406457.2516?

OPERATION.

			40,64,57.25,16(637.54. *Ans.*
			36
Trial divisor,		120	464
Complete divisor,		123	369
Trial	"	1260	9557
Complete	"	1267	8869
Trial	"	1274.0	688.25
Complete	"	1274.5	637.25
Trial	"	1275 00	51.0016
Complete	"	1275.04	51.0016

What is the square root of 2?

			2.	(1.4142+. *Ans.*
			1	
Trial divisor,		20	100	
Complete divisor,		24	96	
Trial	"	280	400	
Complete	"	281	281	
Trial	"	2820	11900	
Complete	"	2824	11296	
Trial	"	28280	60400	
Complete	"	28282	56564	

Application of Square Root.

A Triangle is a figure having three sides and three angles or corners.

A Right-angled Triangle is a figure having three sides and three angles, one of which is a right angle.

In every right-angled triangle, the square of the hypothenuse is equal to the sum of the squares of the base and perpendicular.

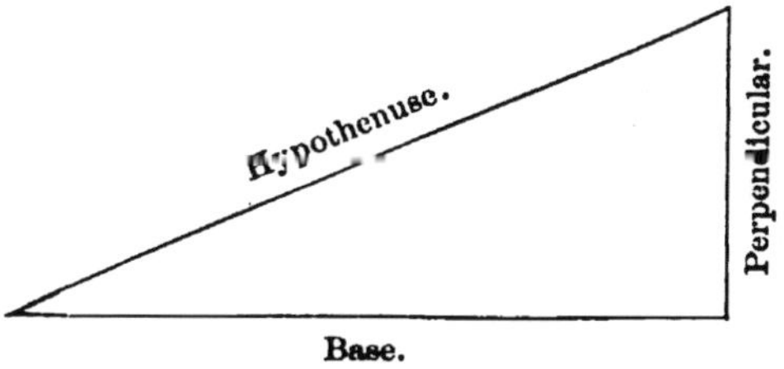

Given the base and perpendicular, to find the hypothenuse.

Rule. — Add the square of the base to the square of the perpendicular, and extract the square root of the sum: the result is the hypothenuse.

Given one side and the hypothenuse, to find the other side.

Rule. — Subtract the square of the given side from the

square of the hypothenuse, and extract the square root of the remainder: the result will be the other side.

Examples.

1. Measure off on the end sill 6 feet from the corner of the house, and on the side sill 8 feet from the same corner: what must be the length of a pole that shall just reach the outside of the sills at those points, when the sills are square? Ans. — 10 feet.

The square of one side is $6 \times 6 = 36$; the square of the other side is $8 \times 8 = 64 + 36 = 100$, the square root of which is 10.

2. A brace has a run of 4 feet × 3 feet 6 inches. What is the length of the brace?

Reduce the feet and inches to inches, in this case; square the length of each run and extract the square root of their sum: the result will be the length of the brace in inches.

3. A square measures 6 feet on a side. What will be the diameter of a circle that shall just enclose it?

The diagonal of the square will be the diameter of the circle.

All circles are to each other as the squares of their radii, diameters, or circumferences.

To find the diameter or circumference of a circle which shall contain a certain number of times the area of a given circle: —

Rule. — Square the given diameter or circumference, and state the question as in proportion; and the fourth term is the square of the required answer, extracting the square root of which gives the answer.

Examples.

1. If a one-inch rope will sustain a weight of 500 lbs., how much will a two-inch rope sustain? $1 \times 1 : 2 \times 2 :: 500$ lbs.: (answer). Ans. — 2,000 lbs.

2. If a $\frac{3}{4}$-inch pipe will empty a cistern in 1 hour 17 minutes, how long will it take a $1\frac{1}{2}$-inch pipe to do it? $\frac{3}{2} \times \frac{3}{2} : \frac{3}{4} \times \frac{3}{4} :: 77$ minutes : (answer). Ans. — $19\frac{1}{4}$ minutes.

3. If a one-inch rope will sustain 500 lbs., what is the size of a rope to sustain 1,000 lbs.? $500 : 1{,}000 :: 1 \times 1 :$ (the square of the answer) $= 2$, the square root of which is $1\frac{3}{8}+$. Ans. — $1\frac{3}{8}+$ inches.

4. If a chain made of $\frac{1}{4}$-inch round iron will sustain a weight of $1\frac{1}{2}$ tons, of what sized iron should a chain be made to sustain a weight of 3 tons? $1\frac{1}{2} : 3 :: \frac{1}{4} \times \frac{1}{4} :$ (the square of the answer) $= \frac{1}{8}$, the square root of which is $.353+ =$ almost $\frac{3}{8}$ inch: therefore a chain made of $\frac{3}{8}$-inch round iron is rather more than twice as strong as one made of $\frac{1}{4}$-inch iron.

TABLE OF SQUARE ROOTS FROM 1 TO 100, INCLUSIVE.

Number.	Square Root.	Number.	Square Root.	Number.	Square Root.	Number.	Square Root.
1	1.0	26	5.09902	51	7.14143	76	8.71779
2	1.41421	27	5.19615	52	7.2111	77	8.77496
3	1.73205	28	5.2915	53	7.28011	78	8.83176
4	2.0	29	5.38517	54	7.34847	79	8.88819
5	2.23607	30	5.47723	55	7.4162	80	8.944
6	2.44948	31	5.56776	56	7.48332	81	9.0
7	2.64575	32	5.65685	57	7.54983	82	9.05538
8	2.82843	33	5.74456	58	7.61577	83	9.11043
9	3.0	34	5.83095	59	7.68115	84	9.16515
10	3.16228	35	5.91608	60	7.74597	85	9.21955
11	3.31663	36	6.0	61	7.81025	86	9.27362
12	3.4641	37	6.08276	62	7.87401	87	9.32738
13	3.60555	38	6.16441	63	7.93725	88	9.38083
14	3.74166	39	6.245	64	8.0	89	9.43398
15	3.87298	40	6.32456	65	8.06226	90	9.48683
16	4.0	41	6.40312	66	8.12404	91	9.53939
17	4.12311	42	6.48074	67	8.18535	92	9.59166
18	4.24264	43	6.55744	68	8.24621	93	9.64365
19	4.3589	44	6.63325	69	8.30662	94	9.69536
20	4.47214	45	6.7082	70	8.3666	95	9.74679
21	4.58258	46	6.7823	71	8.42615	96	9.79796
22	4.69042	47	6.85566	72	8.48528	97	9.84886
23	4.79583	48	6.9282	73	8.544	98	9.89949
24	4.89898	49	7.0	74	8.60233	99	9.94987
25	5.0	50	7.07107	75	8.66025	100	10.0

Cube Root.

The Cube Root is the root of a third power: it is called cube root because the cube or third power of any number represents the contents of a cubic body of which the cube root is the length or breadth of one of the sides.

Rule for extracting the Cube Root. — 1. Point off the given number into periods of three figures each, counting from units' place toward the left in whole numbers, and toward the right in decimals.

2. Find the greatest cube in the left-hand period, and write its root for the first figure in the required root; subtract the cube from the left-hand period, and to the remainder bring down the next period for a dividend.

3. At the left of the dividend write three times the square of the first figure of the root, and annex two ciphers for a trial divisor; divide the dividend by the trial divisor, and write the quotient for a trial figure in the root.

4. Annex the trial figure to three times the former figure, and write the result in a column marked 1, one line below the trial divisor; multiply this term by the trial figure, and write the product on the same line in a column marked 2; add this term as a correction to the trial divisor, and the result will be the complete divisor.

5. Multiply the complete divisor by the trial figure, and subtract the product from the dividend; and to the remainder bring down the next period for a new dividend.

6. Add the square of the last figure of the root, the last term in column 2, and the complete divisor together, and annex two ciphers for a new trial divisor, with which obtain another trial figure in the root.

7. Multiply the unit figure of the last term in column 1 by 3, and annex the trial figure of the root, for the next term of column 2; add this term to the trial divisor for a complete divisor, with which proceed as before.

Note 1. — If at any time the product be greater than the dividend, diminish the trial figure of the root, and correct the erroneous work.

Note 2. — If a cipher occur in the root, annex two more ciphers to the trial divisor, and another period to the dividend; then proceed as before with column 1, annexing both ciphers and trial figure.

EXAMPLE. — What is the cube root of 79.112?

OPERATION.

			79.112 (4 2928+. *Ans.*
			64.
No. 1.	No. 2.	4800	15112
122	244	5044	10088
		529200	5024000
1269	11421	540621	4865589
		55212300	158411000
12872	25744	55238044	110476088
		5526379200	47934912000
128768	1030144	5527409344	44219274752
			3714637248 rem.

Application of the Cube Root.

I wish to make a box, the length, breadth, and depth of which are to be equal, to hold 50 bushels of grain. What is the length of one side of this box?

We first find the number of cubic inches in 50 bushels, then extract the cube root: the result is the length or depth of the box in inches.

Cubes are to each other as the cubes of their sides.

Spheres (round balls) are to each other as the cubes of their diameters or circumferences.

To find the side, diameter, circumference, or altitude of any solid which is similar to a given solid: —

Rule. — State the question as in proportion, and cube the given sides, diameters, circumferences, or altitudes: the cube root of the fourth term of the proportion is the required answer.

Example. — If a two-inch ball weighs 2 pounds, what is the diameter of a ball that weighs twice that?

2 pounds: 4 pounds:: 2 × 2 × 2 inches: (the cube of the answer).

To find the cubical contents or weight of any solid which is similar to a given solid.

Rule. — State the question as in proportion, and cube the given sides, diameters, circumferences, or altitudes: the fourth term of the proportion is the required answer.

Examples. — If a ball 4 inches in diameter weighs 50 pounds, what is the weight of a ball 6 inches in diameter?

4 × 4 × 4: 6 × 6 × 6:: 50 pounds: (the answer).

If a three-inch cube weighs 7 pounds, what is the weight of a four-inch cube?

3 × 3 × 3: 4 × 4 × 4:: 7 pounds: (the answer).

Mensuration.

To find the Area of a Square or Parallelogram. — Multiply the length by the breadth.

To find the Area of a Tapering Board. — Multiply the length in feet, by the breadth of the middle in inches, and divide by 12; or add together the width of the ends in inches, and multiply the length by half of this sum, and divide by 12: the result is the number of square feet contained in the board.

To find the Area of a Rhombus or Rhomboid. — Multiply the length of the side, by the breadth measured square across.

To find the Area of any Triangle. — Multiply the base by half of the perpendicular, or multiply half the base by the perpendicular

To find the Area of a Circle. — Multiply the square of the diameter by .7854.

To find the Circumference of a Circle. — Multiply the diameter by 3.1416.

To find the Surface Area of a Globe. — Multiply the circumference by the diameter.

To find the Solid Contents of a Globe. — Multiply the surface area by $\frac{1}{6}$ of the diameter.

To find the Area of a Ring. — Multiply the sum of the inside and the outside diameters by their difference, and multiply the product thus obtained by .7854.

To find the Side of a Square containing the Same Area as a Given Circle. — Multiply the diameter by .886227.

To find the Side of an Inscribed Square. — Multiply the diameter by .707.

To find the Area of an Ellipse. — Multiply the longer diameter by the shorter, and multiply this product by .7854.

To find the Solid Contents of a Cylinder (as a log). — Multiply the area of the end by the length.

To find the Solid Contents of Pyramids or Cones. — Multiply the area of the base by $\frac{1}{3}$ of the height.

To find the Cubical Contents of the Frustum of a Cone (practical application, find the cubical contents of a tapering, round log). — Multiply together the diameters of the large and of the small ends, and to the product add $\frac{1}{3}$ of the square of the difference of the diameters; then multiply this sum by .7854, which will give the average area; multiply this area by the length, and the product will be the cubical contents.

CIRCLES.

Diam.	Circumf.	Area.	Diam.	Circumf.	Area.	Diam.	Circumf.	Area.
1/64	.049	.00019	4 3/4	14.92	17.72	14 3/4	46.33	170.87
1/32	.0981	.00076	5	15.7	19.635	15	47.12	176.71
1/16	.1963	.00306	5 1/4	16.49	21.647	15 1/4	47.9	182.65
1/8	.3927	.01227	5 1/2	17.27	23.758	15 1/2	48.69	188.69
3/16	.589	.02761	5 3/4	18.06	25.967	15 3/4	49.48	194.82
1/4	.7854	.04908	6	18.84	28.274	16	50.26	201.06
5/16	.9817	.07669	6 1/4	19.63	30.679	16 1/4	51.05	207.39
3/8	1.178	.1104	6 1/2	20.42	33.183	16 1/2	51.83	213.82
7/16	1.374	.1503	6 3/4	21.2	35.784	16 3/4	52.62	220.35
1/2	1.57	.1963	7	21.99	38.484	17	53.4	226.98
9/16	1.767	.2485	7 1/4	22.77	41.282	17 1/4	54.19	233.7
5/8	1.963	.3067	7 1/2	23.56	44.178	17 1/2	54.97	240.52
11/16	2.159	.3712	7 3/4	24.34	47.173	17 3/4	55.76	247.45
3/4	2.356	.4417	8	25.13	50.265	18	56 54	254.46
13/16	2.552	.5184	8 1/4	25.91	53.456	18 1/4	57.33	261.58
7/8	2.748	.6013	8 1/2	26.7	56.745	18 1/2	58.11	268.8
15/16	2.945	.6902	8 3/4	27.48	60.132	18 3/4	58.9	276.11
1	3.1416	.7854	9	28.27	63.617	19	59.69	283.52
1 1/8	3.534	.994	9 1/4	29.05	67.2	19 1/4	60.47	291.03
1 1/4	3.927	1.227	9 1/2	29.84	70.882	19 1/2	61.26	298.64
1 3/8	4.319	1.484	9 3/4	30.63	74 662	19 3/4	62.04	306.35
1 1/2	4.712	1.767	10	31.41	78.539	20	62.83	314.16
1 5/8	5.105	2.073	10 1/4	32.2	82.516	20 1/4	63.61	322.06
1 3/4	5.497	2.405	10 1/2	32.98	86.59	20 1/2	64.4	330.06
1 7/8	5.89	2.761	10 3/4	33.77	90.762	20 3/4	65.18	338.16
2	6.283	3.141	11	34.55	95.033	21	65.97	346.36
2 1/8	6.675	3.546	11 1/4	35.34	99.402	21 1/4	66.75	354.65
2 1/4	7.068	3.976	11 1/2	36.12	103.86	21 1/2	67.54	363.05
2 3/8	7.461	4.43	11 3/4	36.91	108.43	21 3/4	68.32	371.54
2 1/2	7.854	4.908	12	37.69	113.09	22	69.11	380.13
2 5/8	8.246	5.411	12 1/4	38.48	117.85	22 1/4	69.9	388.82
2 3/4	8.639	5.939	12 1/2	39.27	122.71	22 1/2	70.68	397.6
2 7/8	9.032	6.491	12 3/4	40.05	127.67	22 3/4	71.47	406.49
3	9.424	7.068	13	40.84	132.73	23	72.25	415.47
3 1/4	10.21	8.295	13 1/4	41.62	137.88	23 1/4	73.04	424.55
3 1/2	10.99	9.621	13 1/2	42.41	143.13	23 1/2	73.82	433.73
3 3/4	11.78	11.044	13 3/4	43.19	148.48	23 3/4	74.61	443.01
4	12.56	12.566	14	43.98	153 93	24	75.39	452 39
4 1/4	13.35	14.186	14 1/4	44.76	159.48			
4 1/2	14 13	15.904	14 1/2	45.55	165.13			

DECIMAL PARTS OF INCHES.

Dec.	*Frac.*	*Dec.*	*Frac.*
.03125	$\frac{1}{32}$	.53125	$\frac{1}{2}$ $\frac{1}{32}$
.0625	$\frac{1}{16}$	.5625	$\frac{1}{2}$ $\frac{1}{16}$
.09375	$\frac{3}{32}$	.59375	$\frac{1}{2}$ $\frac{3}{32}$
.125	$\frac{1}{8}$	.625	$\frac{5}{8}$
.15625	$\frac{1}{8}$ $\frac{1}{32}$	.65625	$\frac{5}{8}$ $\frac{1}{32}$
.1875	$\frac{1}{8}$ $\frac{1}{16}$	.6875	$\frac{5}{8}$ $\frac{1}{16}$
.21875	$\frac{1}{8}$ $\frac{3}{32}$	.71875	$\frac{5}{8}$ $\frac{3}{32}$
.25	$\frac{1}{4}$	.75	$\frac{3}{4}$
.28125	$\frac{1}{4}$ $\frac{1}{32}$	.78125	$\frac{3}{4}$ $\frac{1}{32}$
.3125	$\frac{1}{4}$ $\frac{1}{16}$	.8125	$\frac{3}{4}$ $\frac{1}{16}$
.34375	$\frac{1}{4}$ $\frac{3}{32}$	.84375	$\frac{3}{4}$ $\frac{3}{32}$
.375	$\frac{3}{8}$	.875	$\frac{7}{8}$
.40625	$\frac{3}{8}$ $\frac{1}{32}$	.90625	$\frac{7}{8}$ $\frac{1}{32}$
.4375	$\frac{3}{8}$ $\frac{1}{16}$	.9375	$\frac{7}{8}$ $\frac{1}{16}$
.46875	$\frac{3}{8}$ $\frac{3}{32}$	.96875	$\frac{7}{8}$ $\frac{3}{32}$
.5	$\frac{1}{2}$	1.	1.

DECIMALS OF FEET.

Dec.	*Frac.*	*Inches.*
.01041	$\frac{1}{96}$	$\frac{1}{8}$ inch.
.02083	$\frac{1}{48}$	$\frac{1}{4}$ "
.03125	$\frac{1}{32}$	$\frac{3}{8}$ "
.04166	$\frac{1}{24}$	$\frac{1}{2}$ "
.05208	$\frac{5}{96}$	$\frac{5}{8}$ "
.0625	$\frac{1}{16}$	$\frac{3}{4}$ "
.07291	$\frac{7}{96}$	$\frac{7}{8}$ "
.0833	$\frac{1}{12}$	1 "
.1666	$\frac{1}{6}$	2 inches.
.25	$\frac{1}{4}$	3 "
.3333	$\frac{1}{3}$	4 "
.4166	$\frac{5}{12}$	5 "
.5	$\frac{1}{2}$	6 "
.5833	$\frac{7}{12}$	7 "
.6666	$\frac{2}{3}$	8 "
.75	$\frac{3}{4}$	9 "
.8333	$\frac{5}{6}$	10 "
.9166	$\frac{11}{12}$	11 "
1.	1.	12 "

THE METRIC SYSTEM. TABLES AUTHORIZED BY CONGRESS.

MEASURES OF LENGTH.

Metric Denominations and Values.		Equivalents in Denominations in Use.
Myriametre	10,000 metres	6.2137 miles.
Kilometre	1,000 metres	0.62137 mile, or 3,280 feet 10 inches.
Hectometre	100 metres	328 feet and 1 inch.
Decametre	10 metres	393.7 inches.
METRE	1 metre	39.37 inches.
Decimetre	$\frac{1}{10}$ of a metre	3.937 inches.
Centimetre........	$\frac{1}{100}$ of a metre	0.3937 inch.
Millimetre	$\frac{1}{1000}$ of a metre	0.0394 inch.

MEASURES OF SURFACES.

Metric Denominations and Values.		Equivalents in Denominations in Use.
Hectare	10,000 square metres......	2.471 acres.
Are	100 square metres......	119.6 square yards.
CENTARE	1 square metre.......	1550 square inches.

MEASURES OF CAPACITY.

Metric Denominations and Values.			Equivalents in Denominations in Use.	
Names.	No. of litres.	Cubic Measure.	Dry Measure.	Liquid or Wine Measure.
Kilolitre, or stere,	1,000	1 cubic metre	1.308 cu. yd..	264.17 gallons.
Hectolitre	100	$\frac{1}{10}$ of a cubic metre ...	2 bu. 3.35 pk.	26.417 gallons.
Decalitre........	10	10 cubic decimetres ..	9.08 quarts ..	2.6417 gallons.
LITRE	1	1 cubic decimetre	0.908 quart ..	1.0567 quarts.
Decilitre	$\frac{1}{10}$	$\frac{1}{10}$ of a cubic decimetre,	6.1022 cu. in..	0.845 gill.
Centilitre	$\frac{1}{100}$	10 cubic centimetres..	0.6102 cu. in..	0.338 fluid oz.
Millilitre	$\frac{1}{1000}$	1 cubic centimetre....	0.061 cu. in...	0.27 fluid dr.

(OVER)

WEIGHTS.

Metric Denominations and Values.			Equivalents in Denominations in Use.
Names.	**No. of grammes.**	**Weight of what quantity of water at maximum density.**	**Avoirdupois Weight.**
Millier, or tonneau..	1,000,000	1 cubic metre	2204.6 pounds.
Quintal	100,000	1 hectolitre	220.46 pounds.
Myriagramme.......	10,000	10 litres....................	22.046 pounds.
Kilogramme, or kilo,	1,000	1 litre......................	2.2046 pounds.
Hectogramme.......	100	1 decilitre	3.5274 ounces.
Decagramme........	10	10 cubic centimetres	0.3527 ounce.
GRAMME............	1	1 cubic centimetre	15.432 grains.
Decigramme	$\frac{1}{10}$	$\frac{1}{10}$ of a cubic centimetre ...	1.5432 grain.
Centigramme	$\frac{1}{100}$	10 cubic millimetres...	0.1543 grain.
Milligramme	$\frac{1}{1000}$	1 cubic millimetre.........	0.0154 grain.

GLOSSARY OF TERMS USED IN ARCHITECTURE AND CARPENTRY.

Acanthus. — An ornament resembling the foliage or leaves of the acanthus plant, used in the capitals of the Corinthian and Composite orders of architecture.

Abutment. — That on which a thing rests, or by which it is supported, as the abutment of an arch.

Arcade. — A series of arches supported by columns or piers, either open or backed by masonry. A long, arched building.

Arris. — The edge formed by two surfaces meeting each other, applied particularly to the raised edges which separate the flutings in a Doric column.

Arris Fillet. — A triangular piece of wood used to raise the covering of a roof against a chimney or wall so as to throw off the rain.

Abacus. — The upper plate upon the capital of a column supporting the architrave.

Architrave. — The lower division of an entablature, or that part which rests immediately on the column. The ornamental moulding around the exterior of an arch. This term is also applied to door and window-casings.

Annulet. — A small, flat fillet encircling a column. It is several times repeated under the Doric capital.

Arch. — A construction of stone or brick arranged in the form of a curve, supporting each other by their mutual pressure.

Astragal. — A little round moulding, which surrounds the top or bottom of a column.

Back-flaps. — Rather long, square hinges, considerably shorter than strap-hinges, but applied in the same manner.

Baluster. — A small column used to support a rail.

Balustrade. — A row of balusters topped by a rail, serving as a fence for balconies, stairs, etc.

Balcony. — A platform projecting from the outside walls of a house, generally enclosed by a balustrade.

Baldachin. — A structure in the form of a canopy, supported by columns or projecting from the wall, placed over doors, thrones, etc.

Band. — A low, flat moulding, broad, but not deep.

Bartizan. — The small, overhanging turret which projects from the angles of towers and other parts of a building.

Base. — The lower projecting part of a room, consisting of the plinth and its mouldings. The part of a column between the top of the pedestal and the bottom of the shaft.

Baston. — A round moulding in the base of a column; also called *Torus.*

Battlement. — A notched parapet; originally used only on fortifications, but since used on buildings.

Batten. — A narrow strip of board used to cover seams or joints in boarding. Any narrow strip of board.

Bay-window. — A window projecting outward from the wall, either in a rectangular, polygonal, or semi-circular form. Sometimes called *Bow-window.*

Bead. — A round moulding. When it comes flush with the surrounding surface, it is called a *Quirk-bead*. when it is raised, it is called a *Cock-bead.* There is also the *Plastering-bead,* which is nailed on to the corner of the stud or furring which forms the external angle of a partition. It is sometimes called a *Rule Joint-bead.*

Beam. — A horizontal timber used to resist a force or weight, as a *Tie-beam*, where it is used to tie the work together; as a *Collar-beam*, when it is used to connect and brace two opposite rafters.

Blockings. — Small pieces of wood fitted and glued in the internal angle formed by the side of one board being fastened to the edge of another, and used to give strength to the joint.

To *Break Joints.* — To arrange the work so that no joint of any course shall come opposite a joint in either the course next above or below it, as seen in shingling, clapboarding, slating, etc.

Butt-joint. — A joint formed by the meeting of the square ends of two pieces of wood, or the joint formed by the square end of one piece meeting the side or edge of another piece.

Bracket. — A piece of wood, stone, or metal projecting from a wall to support shelves, statuary, etc.

Buttress. — A projecting support to the exterior of a wall; most commonly applied to churches in the Gothic style, but also to other buildings, and sometimes to mere walls.

Flying Buttress. — A contrivance for strengthening a part of a building which rises considerably above the rest, consisting of a curved brace or half-arch between it and the opposite face of some lower part, so named from its passing through the air.

Carriage of a stair, also called *Stringer.* — The timber which supports the steps and risers.

Canopy. — An ornamental projection in the Gothic style over doors and windows.

Cantilever. — A projecting block or bracket for supporting a balcony, the upper member of a cornice, the eaves of a house, etc.

Capital. — The header or uppermost part of a column, pilaster, etc. There are six varieties, each adapted to its respective order; viz., the *Gothic*, which is ornamented with leaves and foliations; the *Composite*, also called the *Roman* or *Italic*, which is a combination of the *Ionic* and *Corinthian;* the *Tuscan*, which is plain and unornamented, much resembling the *Doric;* the *Corinthian,* which is distinguished by its profusion of ornaments; the *Doric,* which much resembles the *Tuscan,* and is between that and the *Ionic* in ornamentation; and the *Ionic,* whose distinguishing feature is the volute of its capital, and is less ornamented than the *Corinthian.*

Caul. — A piece of board used in veneering for the purpose of clamping the veneer to the surface to which it is to be glued.

Casement. — A glazed sash or frame which opens on hinges.

Castellated. — Adorned with turrets and battlements like a castle.

Catherine-wheel Window. — An ornamental circular window with radiating divisions or spokes.

Casting or *Warping.* — The bending of a board widthways, caused either by one side shrinking or swelling more than the other, or by the peculiar grain of the wood.

To *Chamfer.* — To bevel the corner of a square-edged piece of wood.

Clamp. — A tool having a screw, used to force and hold work together.

Crown-post. — The middle post of a trussed roof; also called the *King-post.*

Cavetto. — A hollow moulding, whose profile is the quarter of a circle; used chiefly in cornices.

Chancel. — That part of a church between the altar and the rail that encloses it.

Chaptrel. — The capital of a pier or pilaster which receives an arch; also called an *Impost.*

Ceiling. — The upper interior surface opposite the floor.

Choir. — That part of a church appropriated to the use of the officiating clergyman; the *chancel.*

Chord of an arch; the *span.*

Column. — A cylindrical support for roofs, ceilings, etc., composed of base, shaft, and capital; a *pillar.*

Cyma. — A moulding of a cornice which is composed of two members, — a hollow and a round; an ogee moulding. It is called *Cyma Recta* when the upper member is hollow and the lower member is round. It is called *Cyma Reversa* when the upper member is round and the lower member is hollow.

Cinque-foil. — An ornamental foliation, having five points or cusps, used in windows, panels, etc.

Clustered Column. — A column which is composed, or appears to be composed, of several columns collected together.

Composite Order. — An order of architecture made up of the *Ionic* order grafted on the Corinthian; also called the *Roman* or *Italic* order.

Console. — A bracket or shoulder-piece, or a projecting ornament on the keystone of an arch, and often used to support little cornices, busts, and vases.

Coping. — The highest or covering course of masonry in a wall, sometimes bevelled on the top to carry off the water; also called *Capping.*

Concrete. — A mass of stone chippings, pebbles, etc., cemented by mortar, and used for foundations where the soil is light or wet; also used to lay cellar-bottoms. In concrete sidewalks the pebbles are generally cemented by gas-tar instead of mortar or cement.

Cornice. — Any moulded projection which crowns or finishes the part to which it is affixed, as the cornice of an order, of a pediment, of a door, window, or house.

Corinthian Order. — The third order of architecture, characterized by a profusion of ornamentation.

Corbel. — A bracket used to support arches, statuary, etc.

Cove. — An arch overhead where ceilings connect with the walls.

Crocket. — An ornament formed in imitation of curved and bent foliage, and placed upon the angles of spires, canopies, etc.

Curb-roof. — a roof having a double slope; a *gambrel* roof.

Cupola. — A dome-like vault on the top of an edifice, usually on a tower or steeple, as of a public building. The word as commonly used means a small tower or turret built on the top of a building.

Curb-plate. — The plate in a curb-roof that receives the feet of the upper rafters.

Dado. — The die or square part in the middle of the pedestal of a column, between the base and the cornice; also that part of an apartment between the plinth and the impost moulding.

Dentil. — An ornamental square block or projection in cornices, bearing some resemblance to teeth, used particularly in the Ionic, Corinthian, and Composite orders.

Doric Order. — Belonging to the second order of columns, between the Tuscan and Ionic. The Doric order is distinguished for strength and simplicity.

Dormer Window. — A window placed on the inclined plane of the roof of a house, the frame being placed vertically on the rafters.

Echinus. — A moulding of the same form as the ovolo or quarter-round, but properly so called only when ornamented or carved with eggs and anchors.

Engaged Columns. — Columns sunk partly into the wall to which they are attached, and standing out at least one-half of their thickness.

Entablature. — That part of an order which is over the columns, including the architrave, frieze, and cornice.

Façade. — A front view or elevation of an edifice.

Fascia. — A flat member of an order or building, like a flat band or broad fillet.

Festoon. — An ornament of carved work in the form of a wreath of flowers, fruits, and leaves, represented as depending or hanging in an arch.

Fillet. — A little square member or ornament used in various places, but generally as a corona over a greater moulding, sometimes as a small square under other mouldings; also the square part of the *cyma recta* and ogee mouldings.

Finials. — A knot or bunch of foliage that forms the upper extremities of pinnacles in Gothic architecture.

Flute. — A channel in a column or pilaster.

Foil. — A rounded or leaf-like ornament in windows, niches, etc., called trefoil, quatre-foil, quinque-foil, or cinque-foil, etc., according to the number of arcs of which it is composed.

Foliation. — The act of enriching with feather ornaments resembling leaves, or the ornaments themselves.

Frieze. — That part of the entablature of a column which is between the architrave and cornice. It is a flat member or face, often enriched with figures of animals or other ornaments of sculpture, whence its name.

Fresco. — A method of painting on plastered walls and ceilings.

Fret. — An ornament consisting of small fillets intersecting each other at right angles.

Furrings. — Strips of board, 1 by 3 inches, which are nailed on the under side of floor-timbers, to form a level surface for laths, — *strapping.* The term furring is sometimes applied to studding.

Gable. — The vertical triangular end of a house or other building, from the eaves to tho top.

Gable-roof. — The sloping roof which forms a gable.

Gable-window. — A window in a gable, or pointed at the top like a gable.

Girder — The principal piece of timber in a floor, girding or binding the others together.

Gothic. — A style of architecture with high and sharply pointed arches, clustered columns, etc.

Groined Arch. — An arch having an angular curve, made by the intersection of two half-cylinders or arches, as a groined ceiling.

Gutter. — A channel at the eaves of a roof to carry off the rain.

Hanging Buttress. — A buttress supported on a corbel, and not resting on the solid foundation.

Helix. — The little volute under the flowers of the Corinthian capital.

Hip-knob. — An ornament placed upon the roof of a building, either upon the hips or at the point of the gable.

Hip-moulding. — A moulding on the rafter or beam which forms the hip of a building.

Hip-roof. — A roof having sloping ends and sloping sides.

Hood-moulding. — A projecting moulding over the head of an arch.

Interlacing Arches. — Arches usually circular; so constructed that their curves intersect or interlace.

Ionic Order. — An order whose distinguishing feature is the volute of its capital. The column is more slender than the Doric and Tuscan, but less slender and less ornamented than the Corinthian and Composite.

Jamb. — The side-piece or post of a door or window, or any other aperture in a building.

Kerf. — To saw a notch in wood, to make it flexible or easily bent.

King-post. — A post rising from the tie-beam to the roof.

Lattice. — Any work of wood or iron made by crossing laths, rods, or bars, and forming a network.

Lancet Window. — A high and narrow window, pointed like a lancet.

Lintel. — A longitudinal piece of wood or stone placed over a door, window, or other opening; a head-piece.

Louver Window. — An opening in a bell-tower church-steeple, crossed by a series of bars or sloping boards, to exclude the rain, but allow the passage of sound from the bells.

Mantel. — The work over a fireplace in front of a chimney, especially a narrow shelf above the fireplace; called also *Mantel-piece.*

Minaret. — A slender, lofty turret on the mosques of Mohammedan countries, rising by different stages or stories, and surrounded by one or more projecting balconies, from which the people are summoned to prayer.

Mitre. — This term is applied to pieces meeting at an angle, and matching together on a line bisecting the angle: generally, however, an angle of 45° is called a mitre, sometimes called a square mitre; that is to say, a mitre for a square or rectangular figure.

Modillion. — The enriched block or horizontal bracket generally found under the cornice of the Corinthian entablature, and sometimes less ornamented in the Ionic, Composite, and other orders.

Mullion. — A slender bar or pier which forms the divisions between the lights of windows, screens, etc. One of the divisions in panellings resembling windows.

Mutule. — A projecting block worked under the corona of the Doric cornice, in the same situation as the modillion of the Corinthian and Composite orders.

Nave. — The middle or body of a church, extending from the choir or chancel to the principal entrance; also the part between the wings or aisles.

Newel. — The upright post about which the steps of a circular staircase wind; also the principal post at the angles and foot of a staircase.

Niche. — A cavity or recess, generally within the thickness of the wall, for a statue, bust, or other ornament.

Nosing. — That part of the step-board of a stair that projects over the riser; also any like projection.

Ogee. — A moulding consisting of two members, one concave, the other convex, or a round and a hollow.

Oriel Window. — A large bay-window in a hall or chapel.

Ovolo. — A round moulding; the quarter-round.

Pavilion. — A temporary movable building or tent. The name is sometimes given to a summer-house in a garden.

Pedestal. — The base or foot of a column. It consists of three parts, — base, dado or die, and cornice.

Pediment. — The triangular, ornamental facing of a portico, or a similar decoration over doors, windows, gates, etc. The name is also applied to arched and circular ornaments of a like kind.

Pendant. — A hanging ornament, used in roofs, ceilings, etc., much used in Gothic architecture.

Pilaster. — A square column, usually set within a wall, and projecting a fourth or a fifth part of its diameter.

Pinnacle. — A slender turret, or part of a building elevated above the main building.

Pitch of a Roof. — The inclination or slope of the sides; sometimes expressed in parts of the span, as $\frac{1}{3}^{\text{rd}}$ or $\frac{1}{4}^{\text{th}}$ pitch, that is, the rise is that part of the span; sometimes by the length of the rafter in parts of the span, as $\frac{2}{3}^{\text{rds}}$ or $\frac{3}{4}^{\text{ths}}$ pitch, that is, the length of rafter is that part of the span. Also the Gothic pitch, where the length of rafters is the same as the span. Elizabethan pitch, in which the length of rafters is greater than the span. Grecian pitch, in which the rise is $\frac{1}{9}^{\text{th}}$ to $\frac{1}{7}^{\text{th}}$ of the span; and the Roman pitch, in which the rise is $\frac{1}{5}^{\text{th}}$ to $\frac{2}{9}^{\text{ths}}$ of the span.

Plate. — A piece of timber which supports the ends of the rafters.

Plinth. — A square, projecting, vertically faced member, forming the lowest division of the base of a column. The plain, projecting face at the bottom of a wall, immediately above the ground.

Planceer or *Plancher.* — The under side of a cornice; a soffit.

Porch. — A kind of vestibule at the entrance of temples, churches, halls, and other buildings; hence, an ornamental entrance-way.

Portico. — A covered space, enclosed by columns, at the entrance of a building.

Purlin. — A piece of timber extending from end to end of a building or roof, across and under the rafters, to support them in the middle.

Putlog. — A piece of timber on which the planks of a stage are laid, one end resting on the ledger of the stage, and the other in a hole in the wall, left temporarily for the purpose.

Queen-post. — One of the suspended posts in a truss-roof, framed below into the tie-beam, and above into the principal rafter.

Quirk. — A small, acute channel, by which the rounded part of a Grecian ovolo or ogee moulding is separated from the fillet.

Rail. — The horizontal part in any piece of framing or panelling.

Rake. — Pitch or inclination of a roof.

Recess. — Part of a room formed by the receding of the wall, as an alcove, a niche, etc.

Return. — The continuation of a moulding or projection in a different direction.

Seat of a Hip, or *Plan of a Hip.* — A level line over which a hip-rafter stands.

Scotia. — A concave moulding used in the base of a column, between the fillets of the tori, and in other situations. Its outline is a segment of a circle, often greater than a semicircle. The moulding which is put under the nosing of steps.

Scroll. — A convolved or spiral ornament. The volute of the Ionic and Corinthian capitals.

Soffit. — Under side of stairways, archways, entablatures, cornices, or ceilings.

Spire. — A body that shoots up in a conical form; a steeple.

Stall. — A small apartment, where merchandise is exposed for sale, as a butcher's stall.

Stile. — The upright piece in framing or panelling.

Stucco. — Plaster of any kind used as a coating for walls, especially a fine plaster composed of lime or gypsum, with sand and pounded marble; used for internal decoration and fine work.

Surbase. — A cornice or series of mouldings on the top of the base of a pedestal, podium, etc. The surbase of a room is sometimes called a chair-rail.

Tie-beam. — A beam acting as a tie at the bottom of a pair of the principal rafters, and prevents them from thrusting out the walls.

Torus. — A large moulding used in the base of a column. Its profile is semicircular.

Tower. — A lofty building, much higher than it is broad, either standing alone or forming a part of another edifice, — of a church, castle, etc.

Threshold. — The door-sill; the plank, stone, or piece of timber or board, that lies at the bottom or under a door of a house or other building.

Transom. — A horizontal cross-bar over a door or window, sometimes used for the purpose of supporting a sash over a door.

Tracery. — An ornamental divergency of the mullions in the head of a window into arches, curves, and flowing lines, enriched with foliations; the sub-divisions of groined vaults, etc.

Trellis. — A structure or frame of cross-barred work, used for various purposes, as for screens for supporting plants.

Transept. — A part of a church at right angles to the body of the church. In a cruciform church it is one of the arms of the cross.

Turret. — A little tower or spire attached to a building, and rising above it.

Tuscan Order. — The most ancient and simple of the orders of architecture. The capital is plain, unornamental, and much like that of the Doric order.

Veranda. — A kind of open portico, formed by extending a sloping roof beyond the main building.

Vestibule. — The porch or entrance into a house; a hall or ante-chamber next to the entrance, and from which doors open to the various rooms in the house.

Volute. — A kind of spiral scroll used in the Ionic, Corinthian, and Composite orders of architecture.

Well-hole. — The open space in the middle of a staircase beyond the ends of the steps.

Plate 38.

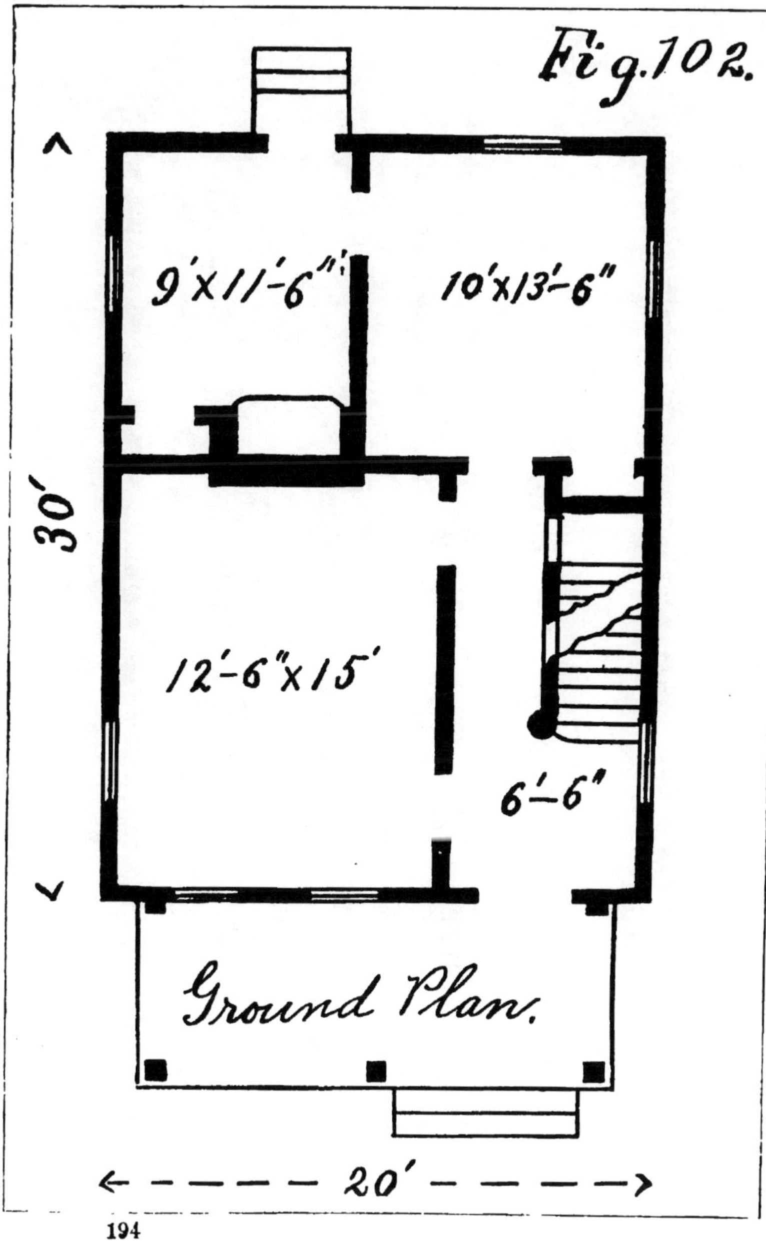

Fig. 103.

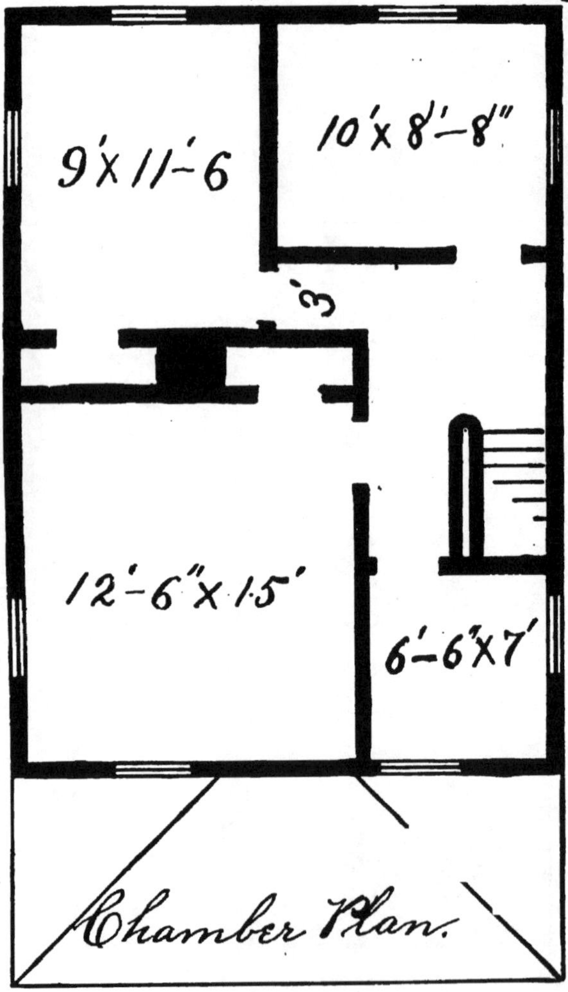

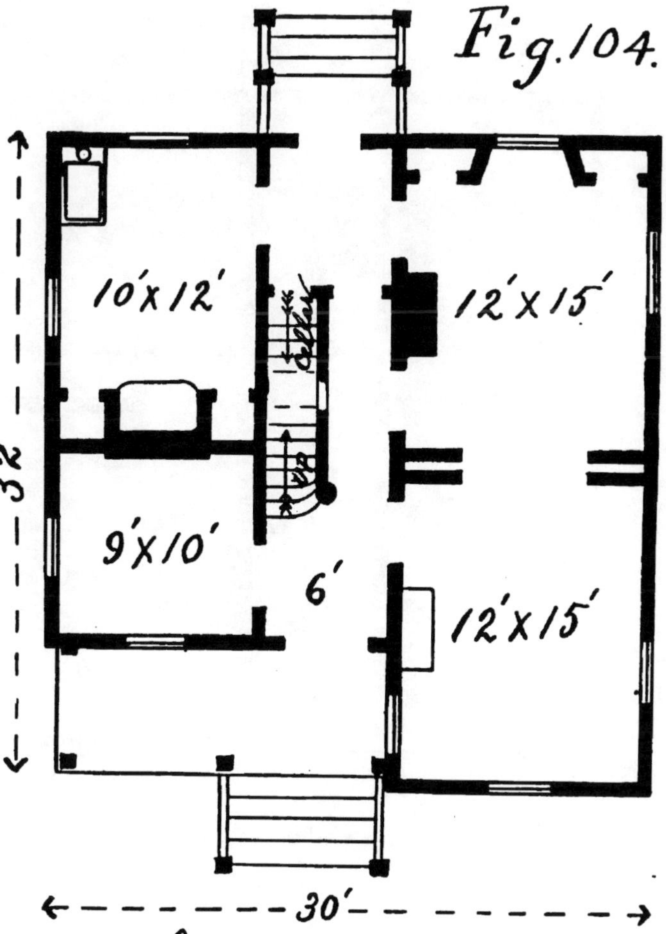

Ground Plan.

Fig. 105.

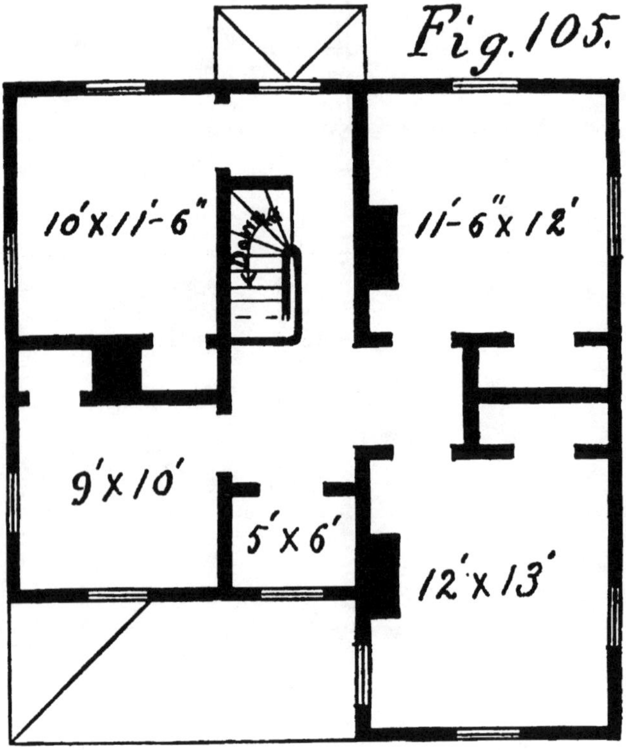

Chamber Plan.

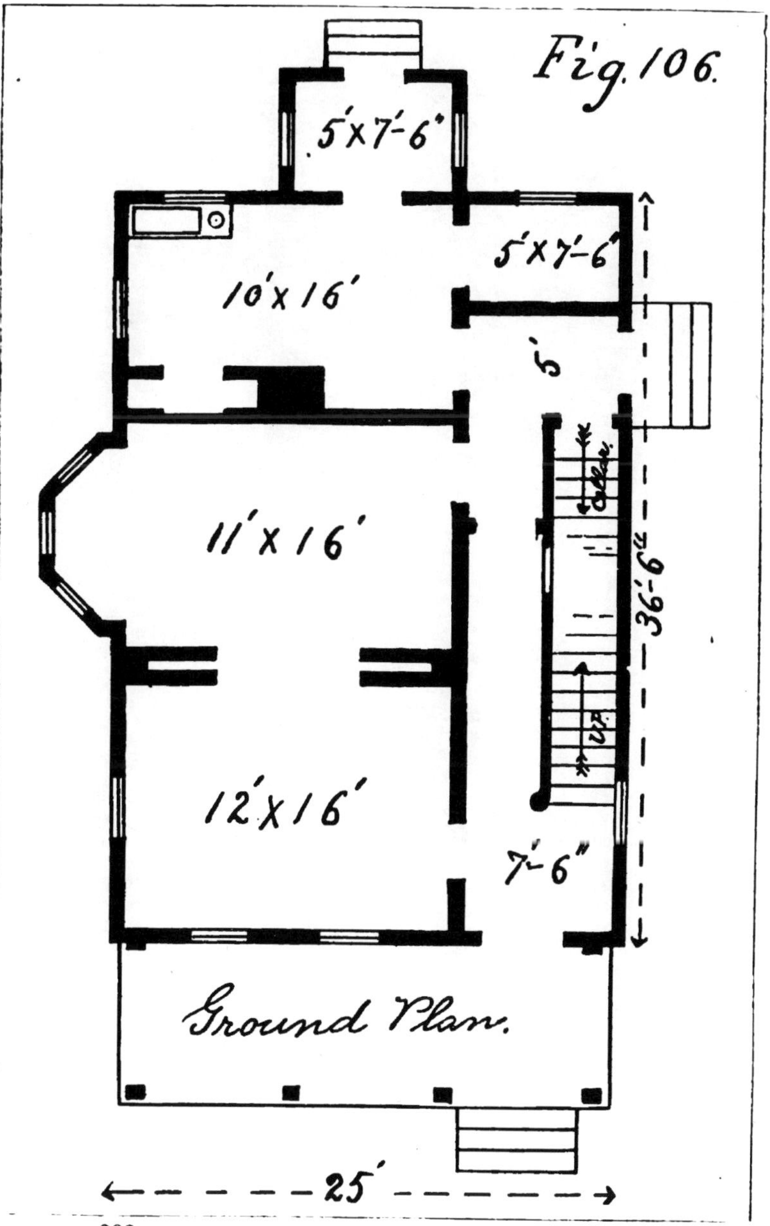
Fig. 106.
5'x7'-6"
10'x16'
5'x7'-6"
5'
11'x16'
Cellar
Up
36'-6"
12'x16'
7'-6"
Ground Plan.
25'

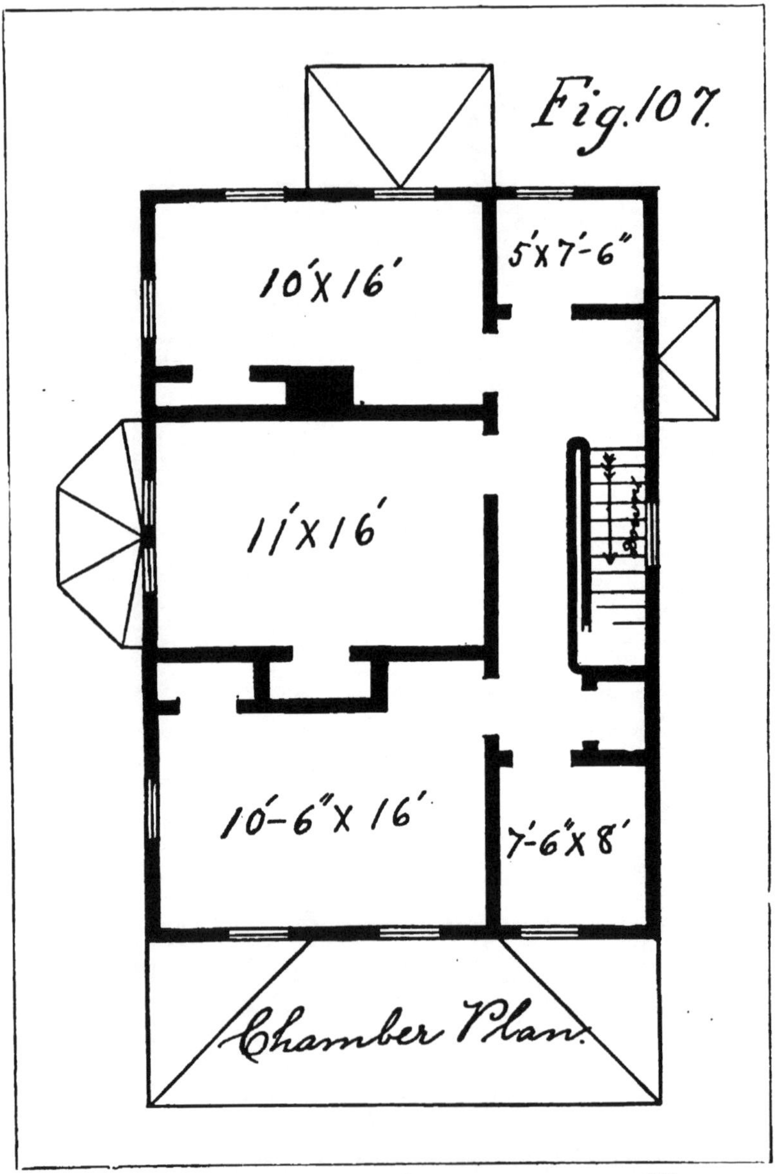
Fig. 107.
10' x 16'
5' x 7'-6"
11' x 16'
Down
10'-6" x 16'
7'-6" x 8'
Chamber Plan.

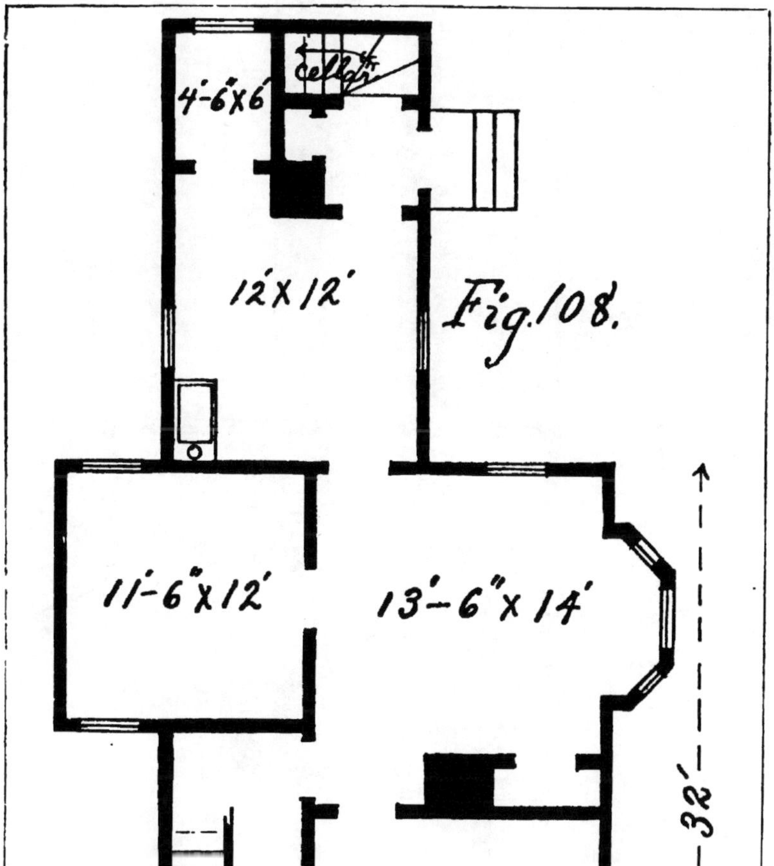

Fig. 108.

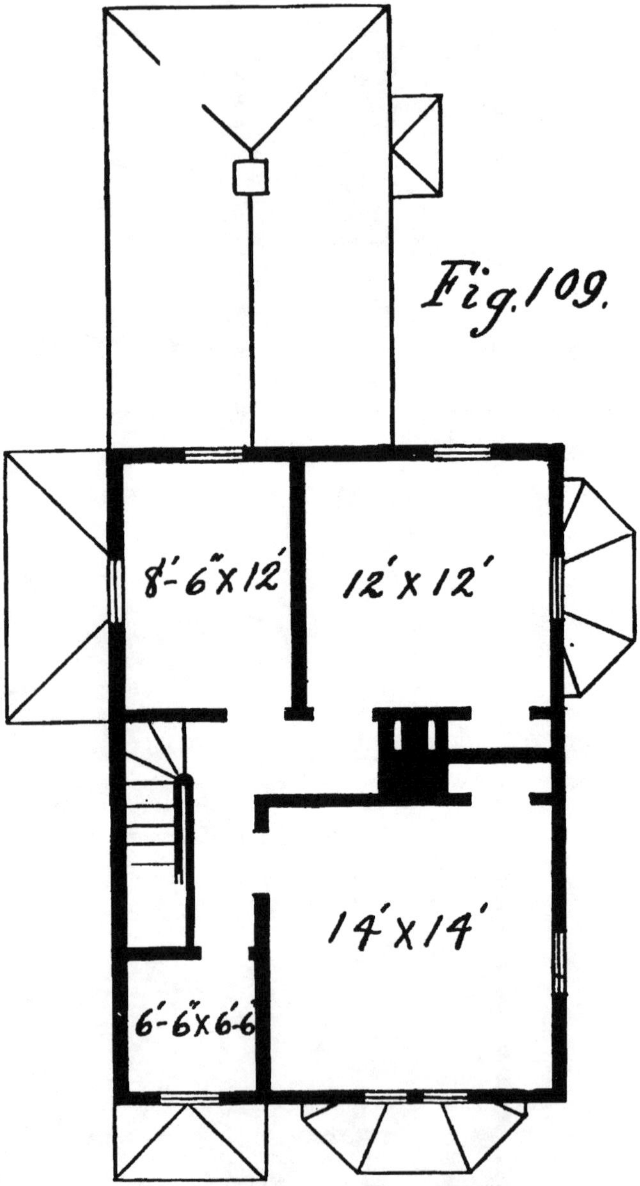

Chamber Plan

INDEX.

Spotted something that isn't quite right?

Let us know and we'll fix it for future reprints. And as a thank you for your diligence, we'll send you a free Sothis book of your choice.

publisher@sothis.press

This page intentionally left blank

Made in the USA
Monee, IL
08 November 2026

68231665R00127